# Algebra I
# Essentials

for
dummies®
A Wiley Brand

# Algebra I
# Essentials

WITHDRAWN

## by Mary Jane Sterling

A Wiley Brand

# Algebra I Essentials For Dummies®

Published by: **John Wiley & Sons, Inc.**, 111 River Street, Hoboken, NJ 07030-5774, www.wiley.com

Copyright © 2019 by John Wiley & Sons, Inc., Hoboken, New Jersey

Published simultaneously in Canada

For general information on our other products and services, please contact our Customer Care Department within the U.S. at 877-762-2974, outside the U.S. at 317-572-3993, or fax 317-572-4002. For technical support, please visit https://hub.wiley.com/community/support/dummies.

Wiley publishes in a variety of print and electronic formats and by print-on-demand. Some material included with standard print versions of this book may not be included in e-books or in print-on-demand. If this book refers to media such as a CD or DVD that is not included in the version you purchased, you may download this material at http://booksupport.wiley.com. For more information about Wiley products, visit www.wiley.com.

Library of Congress Control Number: 2019932871

ISBN: 978-1-119-59096-5 (pbk); ISBN: 978-1-119-59098-9 (ePDF); ISBN: 978-1-119-59095-8 (ePub)

Manufactured in the United States of America

V10009762_042419

# Contents at a Glance

# Table of Contents

## CHAPTER 10: Absolute-Value Equations and Inequalities .......................... 115

## CHAPTER 11: Making Algebra Tell a Story ........................ 121

## CHAPTER 12: Putting Geometry into Story Problems ............... 133

# Introduction

One of the most commonly asked questions in a mathematics classroom is, "What will I ever use this for?" Some teachers can give a good, convincing answer. Others hem and haw and stare at the floor. My favorite answer is, "Algebra gives you power." Algebra gives you the power to move on to bigger and better things in mathematics. Algebra gives you the power of knowing that you know something that your neighbor doesn't know. Algebra gives you the power to be able to help someone else with an algebra task or to explain to your child these logical mathematical processes.

Algebra is a system of symbols and rules that is universally understood, no matter what the spoken language. Algebra provides a clear, methodical process that can be followed from beginning to end. What *power!*

## About This Book

What could be more *essential* than *Algebra I Essentials For Dummies?* In this book, you find the main points, the nitty-gritty (made spiffy-jiffy), and a format that lets you find what you need about an algebraic topic as you need it. I keep the same type of organization that you find in *Algebra I For Dummies,* 2nd Edition, but I keep the details neat, sweet, and don't repeat. The fundamentals are here for your quick reference or, if you prefer, a more thorough perusal. The choice is yours.

This book isn't like a mystery novel; you don't have to read it from beginning to end. I divide the book into some general topics — from the beginning vocabulary and processes and operations to the important tool of factoring to equations and applications. So you can dip into the book wherever you want, to find the information you need.

# Conventions Used in This Book

I don't use many conventions in this book, but you should be aware of the following:

>> When I introduce a new term, I put that term in *italics* and define it nearby (often in parentheses).

>> I express numbers or numerals either with the actual symbol, such as 8, or the written-out word: *eight.* Operations, such as + are either shown as this symbol or written as *plus.* The choice of expression all depends on the situation — and on making it perfectly clear for you.

# Foolish Assumptions

I don't assume that you're as crazy about math as I am — and you may be even *more* excited about it than I am! I do assume, though, that you have a mission here — to brush up on your skills, improve your mind, or just have some fun. I also assume that you have some experience with algebra — full exposure for a year or so, maybe a class you took a long time ago, or even just some preliminary concepts.

You may be delving into the world of algebra again to refresh those long-ago lessons. Is your kid coming home with assignments that are beyond your memory? Are you finally going to take that calculus class that you've been putting off? Never fear. Help is here!

# Icons Used in This Book

The little drawings in the margin of the book are there to draw your attention to specific text. Here are the icons I use in this book:

To make everything work out right, you have to follow the basic rules of algebra (or mathematics in general). You can't change or ignore them and arrive at the right answer. Whenever I give you an algebra rule, I mark it with this icon.

An explanation of an algebraic process is fine, but an example of how the process works is even better. When you see the Example icon, you'll find one or more problems using the topic at hand.

Paragraphs marked with the Remember icon help clarify a symbol or process. I may discuss the topic in another section of the book, or I may just remind you of a basic algebra rule that I discuss earlier.

The Tip icon isn't life-or-death important, but it generally can help make your life easier — at least your life in algebra.

The Warning icon alerts you to something that can be particularly tricky. Errors crop up frequently when working with the processes or topics next to this icon, so I call special attention to the situation so you won't fall into the trap.

# Where to Go from Here

If you want to refresh your basic skills or boost your confidence, start with the fractions, decimals, and signed numbers in the first chapter. Other essential concepts are the exponents in Chapter 2 and order of operations in Chapter 3. If you're ready for some factoring practice and need to pinpoint which method to use with what, go to Chapters 4 and 5. Chapters 6, 7, and 8 are for you if you're ready to solve equations; you can find just about any type you're ready to attack. Chapters 9 and 10 get you back into inequalities and absolute value. And Chapters 11 and 12 are where the good stuff is: applications — things you can do with all those good solutions. I finish with some graphing in Chapter 13 and then give you a list of pitfalls to avoid in Chapter 14.

Studying algebra can give you some logical exercises. As you get older, the more you exercise your brain cells, the more alert and "with it" you remain. "Use it or lose it" means a lot in terms of the brain. What a good place to use it, right here!

The best *why* for studying algebra is just that it's beautiful. Yes, you read that right. Algebra is poetry, deep meaning, and artistic expression. Just look, and you'll find it. Also, don't forget that it gives you *power*.

Welcome to algebra! Enjoy the adventure!

# Beyond the Book

In addition to what you're reading right now, this book comes with a free access-anywhere Cheat Sheet. To get this Cheat Sheet, go to www.dummies.com and search for "Algebra I Essentials For Dummies Cheat Sheet" by using the Search box.

# Chapter **1**

# Setting the Scene for Actions in Algebra

What exactly is algebra? What is it *really* used for? In a nutshell, *algebra* is a systematic study of numbers and their relationships, using specific rules. You use *variables* (letters representing numbers), and formulas or equations involving those variables, to solve problems. The problems may be practical applications, or they may be puzzles for the pure pleasure of solving them!

In this chapter, I acquaint you with the various number systems. You've seen the numbers before, but I give you some specific names used to refer to them properly. I also tell you how I describe the different processes performed in algebra — I want to use the correct language, so I give you the vocabulary. And, finally, I get very specific about fractions and decimals and show you how to move from one type to the other with ease.

## Making Numbers Count

Algebra uses different types of numbers, in different circumstances. The types of numbers are important because what they look like and how they behave can set the scene for particular situations or help to solve particular problems. Sometimes it's really convenient to declare, "I'm only going to look at whole-number answers," because whole numbers do not include fractions or negatives. You could easily end up with a fraction if you're working through a problem that involves a number of cars or people. Who wants half a car or, heaven forbid, a third of a person?

I describe the different types of numbers in the following sections.

## Facing reality with reals

*Real numbers* are just what the name implies: real. Real numbers represent real values — no pretend or make-believe. They cover the gamut and can take on any form — fractions or whole numbers, decimal numbers that go on forever and ever without end, positives and negatives.

## Going green with naturals

A *natural number* (also called a *counting number*) is a number that comes naturally. The natural numbers are the numbers starting with 1 and going up by ones: 1, 2, 3, 4, 5, and so on into infinity.

## Wholesome whole numbers

*Whole numbers* aren't a whole lot different from natural numbers (see the preceding section). Whole numbers are just all the natural numbers plus a 0: 0, 1, 2, 3, 4, 5, and so on into infinity.

## Integrating integers

*Integers* are positive and negative whole numbers: . . . –3, –2, –1, 0, 1, 2, 3, . . . .

Integers are popular in algebra. When you solve a long, complicated problem and come up with an integer, you can be joyous because your answer is probably right. After all, most teachers like answers without fractions.

## Behaving with rationals

*Rational numbers* act rationally because their decimal equivalents behave. The decimal ends somewhere, or it has a repeating pattern to it. That's what constitutes "behaving."

Some rational numbers have decimals that end such as: 3.4, 5.77623, −4.5. Other rational numbers have decimals that repeat the same pattern, such as 3.164164$\overline{164}$, or 0.66666666$\overline{6}$. The horizontal bar over the 164 and the 6 lets you know that these numbers repeat forever.

TIP

In *all* cases, rational numbers can be written as fractions. Each rational number has a fraction that it's equal to. So one definition of a *rational number* is any number that can be written as a fraction, $\frac{p}{q}$, where $p$ and $q$ are integers (except $q$ can't be 0). If a number can't be written as a fraction, then it isn't a rational number.

## Reacting to irrationals

Irrational numbers are just what you may expect from their name — the opposite of rational numbers. An *irrational number* can't be written as a fraction, and decimal values for irrationals never end and never have the same, repeated pattern in them.

## Picking out primes and composites

A number is considered to be *prime* if it can be divided evenly only by 1 and by itself. The first prime numbers are: 2, 3, 5, 7, 11, 13, 17, 19, 23, 29, 31, and so on. The only prime number that's even is 2, the first prime number.

A number is *composite* if it isn't prime — if it can be divided by at least one number other than 1 and itself. So the number 12 is composite because it's divisible by 1, 2, 3, 4, 6, and 12.

# Giving Meaning to Words and Symbols

Algebra and symbols in algebra are like a foreign language. They all mean something and can be translated back and forth as needed. Knowing the vocabulary in a foreign language is important — and it's just as important in algebra.

## Valuing vocabulary

Using the correct word is so important in mathematics. The correct wording is shorter, more descriptive, and has an exact mathematical meaning. Knowing the correct word or words eliminates misinterpretations and confusion.

>> An *expression* is any combination of values and operations that can be used to show how things belong together and compare to one another. An example of an expression is $2x^2 + 4x$.

>> A *term*, such as 4*xy*, is a grouping together of one or more *factors*. Multiplication is the only thing connecting the number with the variables. Addition and subtraction, on the other hand, separate terms from one another, such as in the expression $3xy + 5x - 6$.

>> An *equation* uses a sign to show a relationship — that two things are equal. An example is $2x^2 + 4x = 7$.

>> An *operation* is an action performed upon one or two numbers to produce a resulting number. Operations are addition, subtraction, multiplication, division, square roots, and so on.

>> A *variable* is a letter representing some unknown; a variable always represents a number, but it varies until it's written in an equation or inequality. (An *inequality* is a comparison of two values.) By convention, mathematicians usually assign letters at the end of the alphabet (such as *x*, *y*, and *z*) to be variables.

>> A *constant* is a value or number that never changes in an equation — it's constantly the same. For example, 5 is a constant because it is what it is. By convention, mathematicians usually assign letters at the beginning of the alphabet (such as *a*, *b*, and *c*) to represent constants. In the equation $ax^2 + bx + c = 0$, *a*, *b*, and *c* are constants and *x* is the variable.

>> An *exponent* is a small number written slightly above and to the right of a variable or number, such as the 2 in the expression $3^2$. It's used to show repeated multiplication. An exponent is also called the *power* of the value.

## Signing up for symbols

The basics of algebra involve symbols. Algebra uses symbols for quantities, operations, relations, or grouping. The symbols are shorthand and are much more efficient than writing out the words or meanings.

>> **+** means *add, find the sum, more than,* or *increased by;* the result of addition is the *sum.* It's also used to indicate a *positive number.*

>> **−** means *subtract, minus, decreased by,* or *less than;* the result is the *difference.* It's also used to indicate a *negative number.*

>> **×** means *multiply* or *times.* The values being multiplied together are the *multipliers* or *factors;* the result is the *product.*

**TIP**

In algebra, the × symbol is used infrequently because it can be confused with the variable *x.* You can use · or * in place of × to eliminate confusion.

Some other symbols meaning multiply can be grouping symbols: ( ), [ ], { }. The grouping symbols are used when you need to contain many terms or a messy expression. By themselves, the grouping symbols don't mean to multiply, but if you put a value in front of a grouping symbol, it means to multiply. (See the next section for more on grouping symbols.)

>> **÷** means *divide.* The *divisor* divides the *dividend.* The result is the *quotient.* Other signs that indicate division are the fraction line and the slash (/).

>> $\sqrt{\ }$ means to take the *square root* of something — to find the number that, multiplied by itself, gives you the number under the sign.

>> | | means to find the *absolute value* of a number, which is the number itself (if the number is positive) or its distance from 0 on the number line (if the number is negative).

>> π is the Greek letter pi, which refers to the irrational number: 3.14159. . . . It represents the relationship between the diameter and circumference of a circle: $\pi = \frac{c}{d}$, where *c* is circumference and *d* is diameter.

>> ≈ means *approximately equal* or *about equal.* This symbol is useful when you're rounding a number.

## Going for grouping

In algebra, tasks are accomplished in a particular order. After following the order of operations (see Chapter 3), you have to do what's inside a grouping symbol before you can use the result in the rest of the equation.

*Grouping symbols* tell you that you have to deal with the terms inside the grouping symbols *before* you deal with the larger problem. If the problem contains grouped items, do what's inside a grouping symbol first, and then follow the order of operations. The grouping symbols are

>> **Parentheses ( ):** Parentheses are the most commonly used symbols for grouping.

>> **Brackets [ ] and braces { }:** Brackets and braces are also used frequently for grouping and have the same effect as parentheses.

TIP

Using the different types of grouping symbols helps when there's more than one grouping in a problem. It's easier to tell where a group starts and ends.

>> **Radical $\sqrt{\ }$ :** This symbol is used for finding roots.

>> **Fraction line:** The fraction line also acts as a grouping symbol — everything in the *numerator* (above the line) is grouped together, and everything in the *denominator* (below the line) is grouped together.

# Operating with Signed Numbers

The basic operations are addition, subtraction, multiplication, and division. When you're performing those operations on positive numbers, negative numbers, and mixtures of positive and negative numbers, you need to observe some rules, which I outline in this section.

## Adding signed numbers

You can add positive numbers to positive numbers, negative numbers to negative numbers, or any combination of positive and negative numbers. Let's start with the easiest situation: when the numbers have the same sign.

There's a nice S rule for addition of positives to positives and negatives to negatives. See if you can say it quickly three times in a row: *When the signs are the same, you find the sum, and the sign of the sum is the same as the signs.* This rule holds when $a$ and $b$ represent any two positive real numbers:

$$(+a)+(+b) = +(a+b) \qquad (-a)+(-b) = -(a+b)$$

Here are some examples of finding the sums of same-signed numbers:

>> $(+8)+(+11) = +19$: The signs are all positive.

>> $(-14)+(-100) = -114$: The sign of the sum is the same as the signs.

>> $(+4)+(+7)+(+2) = +13$: Because all the numbers are positive, add them and make the sum positive, too.

>> $(-5)+(-2)+(-3)+(-1) = -11$: This time all the numbers are negative, so add them and give the sum a minus sign.

Numbers with different signs add up very nicely. You just have to know how to do the computation.

When the signs of two numbers are different, forget the signs for a while and find the *difference* between the numbers. This is the difference between their *absolute values*. The number farther from zero determines the sign of the answer:

>> $(+a)+(-b) = +(|a|-|b|)$ if the positive $a$ is farther from zero.

>> $(+a)+(-b) = -(|b|-|a|)$ if the negative $b$ is farther from zero.

Here are some examples of finding the sums of numbers with different signs:

>> $(+6)+(-7) = -1$: The difference between 6 and 7 is 1. Seven is farther from 0 than 6 is, and 7 is negative, so the answer is -1.

>> $(-6)+(+7) = +1$: This time the 7 is positive and the 6 is negative. Seven is still farther from 0 than 6 is, and the answer this time is +1.

# Subtracting signed numbers

Subtracting signed numbers is really easy to do: You *don't!* Instead of inventing a new set of rules for subtracting signed numbers, mathematicians determined that it's easier to change the subtraction problems to addition problems and use the rules I explain in the previous section. But, to make this business of changing a subtraction problem to an addition problem give you the correct answer, you really change *two* things. (It almost seems to fly in the face of *two wrongs don't make a right,* doesn't it?)

**ALGEBRA RULES**

When subtracting signed numbers, change the minus sign to a plus sign *and* change the number that the minus sign was in front of to its opposite. Then just add the numbers using the rules for adding signed numbers:

>> $(+a) - (+b) = (+a) + (-b)$

>> $(+a) - (-b) = (+a) + (+b)$

>> $(-a) - (+b) = (-a) + (-b)$

>> $(-a) - (-b) = (-a) + (+b)$

**EXAMPLE**

Here are some examples of subtracting signed numbers:

>> $-16 - 4 = -16 + (-4) = -20$: The subtraction becomes addition, and the +4 become negative. Then, because you're adding two signed numbers with the same sign, you find the sum and attach their common negative sign.

>> $-3 - (-5) = -3 + (+5) = 2$: The subtraction becomes addition, and the –5 becomes positive. When adding numbers with opposite signs, you find their difference. The 2 is positive, because the +5 is farther from 0.

>> $9 - (-7) = 9 = (+7) = 16$: The subtraction becomes addition, and the –7 becomes positive. When adding numbers with the same sign, you find their sum. The two numbers are now both positive, so the answer is positive.

# Multiplying and dividing signed numbers

Multiplication and division are really the easiest operations to do with signed numbers. As long as you can multiply and divide, the rules are not only simple, but the same for both operations.

**ALGEBRA RULES**

When multiplying and dividing two signed numbers, if the two signs are the same, then the result is *positive;* when the two signs are different, then the result is *negative:*

» $(+a) \cdot (+b) = +ab$

» $(+a) \div (+b) = +(a \div b)$

» $(+a) \cdot (-b) = -ab$

» $(+a) \div (-b) = -(a \div b)$

» $(-a) \cdot (+b) = -ab$

» $(-a) \div (+b) = -(a \div b)$

» $(-a) \cdot (-b) = +ab$

» $(-a) \div (-b) = +(a \div b)$

Notice in which cases the answer is positive and in which cases it's negative. You see that it doesn't matter whether the negative sign comes first or second, when you have a positive and a negative. Also, notice that multiplication and division seem to be "as usual" except for the positive and negative signs.

**EXAMPLE**

Here are some examples of multiplying and dividing signed numbers:

» $(-8) \cdot (+2) = -16$

» $(-5) \cdot (-11) = +55$

» $(+24) \div (-3) = -8$

» $(-30) \div (-2) = +15$

You can mix up these operations doing several multiplications or divisions or a mixture of each and use the following even-odd rule.

**ALGEBRA RULES**

According to the even-odd rule, when multiplying and dividing a bunch of numbers, count the number of negatives to determine the final sign. An *even* number of negatives means the result is *positive.* An *odd* number of negatives means the result is *negative.*

**EXAMPLE**

Here are some examples of multiplying and dividing collections of signed numbers:

» $(+2) \cdot (-3) \cdot (+4) = -24$: This problem has just one negative sign. Because 1 is an odd number (and often the loneliest number), the answer is negative. The numerical parts (the 2, 3, and 4) get multiplied together and the negative is assigned as its sign.

» $(+2) \cdot (-3) \cdot (+4) \cdot (-1) = -24$: Two negative signs mean a positive answer because 2 is an even number.

» $\dfrac{(+4) \cdot (-3)}{(-2)} = +6$: An even number of negatives means you have a positive answer.

» $\dfrac{(-12) \cdot (-6)}{(-4) \cdot (+3)} = -6$: Three negatives yield a negative.

# Dealing with Decimals and Fractions

Numbers written as repeating or terminating decimals have fractional equivalents. Some algebraic situations work better with decimals and some with fractions, so you want to be able to pick and choose the one that's best for your situation.

## Changing fractions to decimals

All fractions can be changed to decimals. Earlier in this chapter, I tell you that rational numbers have decimals that can be written exactly as fractions. The decimal forms of rational numbers either terminate (end) or repeat in a pattern.

**ALGEBRA RULES**

To change a fraction to a decimal, just divide the top by the bottom:

» $\dfrac{7}{4}$ becomes $4\overline{)7.00} = 1.75$, so $\dfrac{7}{4} = 1.75$.

» $\dfrac{4}{11}$ becomes $11\overline{)4.000000\ldots} = 0.363636\ldots$ so $\dfrac{4}{11} = 0.363636\ldots$
= $0.\overline{36}$. The division never ends, so the three dots (ellipses) or bar across the top tell you that the pattern repeats forever.

If the division doesn't come out evenly, you can either show the repeating digits or you can stop after a certain number of decimal places and round off.

## Changing decimals to fractions

Decimals representing rational numbers come in two varieties: terminating decimals and repeating decimals. When changing from decimals to fractions, you put the digits in the decimal over some other digits and reduce the fraction.

### Getting terminal results with terminating decimals

ALGEBRA
RULES

To change a terminating decimal into a fraction, put the digits to the right of the decimal point in the numerator. Put the number 1 in the denominator followed by as many zeros as the numerator has digits. Reduce the fraction if necessary.

EXAMPLE

Change 0.36 into a fraction:

$$0.36 = \frac{36}{100} = \frac{9}{25}$$

There are two digits in 36, so the 1 in the denominator is followed by two zeros. Both 36 and 100 are divisible by 4, so the fraction reduces.

EXAMPLE

Change 0.0005 into a fraction:

$$0.0005 = \frac{5}{10,000} = \frac{1}{2,000}$$

Don't forget to count the zeros in front of the 5 when counting the number of digits. The fraction reduces.

### Repeating yourself with repeating decimals

When a decimal repeats itself, you can always find the fraction that corresponds to the decimal. In this chapter, I only cover the decimals that show every digit repeating.

ALGEBRA
RULES

To change a *repeating decimal* (in which every digit is part of the repeated pattern) into its corresponding fraction, write the repeating digits in the numerator of a fraction and, in the denominator, as many nines as there are repeating digits. Reduce the fraction if necessary.

**EXAMPLE**

Here are some examples of changing the repeating decimals to fractions:

» $0.126126126\ldots = \dfrac{126}{999} = \dfrac{14}{111}$: The three repeating digits are 126. Placing the 126 over a number with three 9s, you reduce by dividing the numerator and denominator by 9.

» $0.857142857142857142\ldots = \dfrac{857{,}142}{999{,}999} = \dfrac{6}{7}$: The six repeating digits are put over six nines. Reducing the fraction takes a few divisions. The common factors of the numerator and denominator are 11, 13, 27, and 37.

# Chapter 2
# Examining Powers and Roots

E xponents were developed so that mathematicians wouldn't have to keep repeating themselves! What is an exponent? An *exponent* is the small, superscripted number to the upper right of the larger number that tells you how many times you multiply the larger number, called the *base*.

## Expanding and Contracting with Exponents

When algebra was first written with symbols — instead of with all words — there were no exponents. If you wanted to multiply the variable $y$ times itself six times, you'd write it: $yyyyyy$. Writing the variable over and over can get tiresome, so the wonderful system of exponents was developed.

The base of an exponential expression can be any real number (see Chapter 1 for more on real numbers). The exponent can be any real number, too, as long as rules involving radicals aren't violated (see "Circling around Square Roots," later in this chapter). An exponent can be positive, negative, fractional, or even a radical. What power!

When a number $x$ is involved in repeated multiplication of $x$ times itself, then the number $n$ can be used to describe how many multiplications are involved: $x^n = x \cdot x \cdot x \cdot x \cdot x \ldots n$ times.

Even though the $x$ in the expression $x^n$ can be any real number and the $n$ can be any real number, they can't both be 0 at the same time. For example, $0^0$ really has no meaning in algebra. It takes a calculus course to prove why this restriction is so. Also, if $x$ is equal to 0, then $n$ can't be negative.

Here are two examples using exponential notation:

>> $2^4 = 2 \cdot 2 \cdot 2 \cdot 2 = 16$

>> $5^{-3} = \dfrac{1}{5^3} = \dfrac{1}{5 \cdot 5 \cdot 5} = \dfrac{1}{125}$

When the exponent is negative, you apply the rule involving rewriting negative exponents before writing the product. (See "Taking on the Negativity of Exponents," later in this chapter.)

# Exhibiting Exponent Products

You can multiply many exponential expressions together without having to change their form into the big or small numbers they represent. The only requirement is that the bases of the exponential expressions that you're multiplying have to be the same. The answer is then a nice, neat exponential expression.

You *can* multiply $2^4 \cdot 2^6$ and $a^5 \cdot a^8$, but you *cannot* multiply $3^6 \cdot 4^7$ using the rule, because the bases are not the same.

To multiply powers of the same base, add the exponents together: $x^a \cdot x^b = x^{a+b}$.

Here are two examples of finding the products of the numbers by adding the exponents:

>> $2^4 \cdot 2^6 = 2^{4+6} = 2^{10}$

>> $a^5 \cdot a^8 = a^{13}$

Often, you find algebraic expressions with a whole string of factors; you want to simplify the expression, if possible. When there's more than one base in an expression with powers, you combine the numbers with the same bases, find the values, and then write them all together.

Here's how to simplify the following expressions:

**EXAMPLE**

» $3^2 \cdot 2^2 \cdot 3^3 \cdot 2^4 = 3^{2+3} \cdot 2^{2+4} = 3^5 \cdot 2^6$: The two factors with base 3 combine, as do the two factors with base 2.

» $4x^6 y^5 x^4 y = 4x^{6+4} y^{5+1} = 4x^{10} y^6$: The number 4 is a coefficient, which is written before the rest of the factors.

When there's no exponent showing, such as with $y$, you assume that the exponent is 1. In the preceding example, you can see that the factor $y$ was written as $y^1$ so its exponent could be added to that in the other $y$ factor.

**REMEMBER**

# Taking Division to Exponents

You can divide exponential expressions, leaving the answers as exponential expressions, as long as the bases are the same. Division is the opposite of multiplication, so it makes sense that, because you add exponents when multiplying numbers with the same base, you *subtract* the exponents when dividing numbers with the same base. Easy enough?

To divide powers with the same base, subtract the exponents: $\frac{x^a}{x^b} = x^a \div x^b = x^{a-b}$, where $x$ can be any real number except 0. (*Remember:* You can't divide by 0.)

**ALGEBRA RULES**

Here are two examples of simplifying expressions by dividing:

**EXAMPLE**

» $2^{10} \div 2^4 = 2^{10-4} = 2^6$: These exponentials represent the equation $1,024 \div 16 = 64$. It's much easier to leave the numbers as bases with exponents.

» $\frac{4x^6 y^3 z^2}{2x^4 y^3 z} = 2x^{6-4} y^{3-3} z^{2-1} = 2x^2 y^0 z^1 = 2x^2 z$

Did you wonder where the $y$ factor went? For more on $y^0$, read on.

# Taking on the Power of Zero

If $x^3$ means $x \cdot x \cdot x$, what does $x^0$ mean? Well, it doesn't mean $x$ times 0, so the answer isn't 0. $x$ represents some unknown real number; real numbers can be raised to the 0 power — except that the base just can't be 0. To understand how this works, use the following rule for division of exponential expressions involving 0.

**ALGEBRA RULES**

Any number to the power of 0 equals 1 as long as the base number is not 0. In other words, $a^0 = 1$ as long as $a \neq 0$.

Here are two examples of simplifying, using the rule that when you raise a real number $a$ to the 0 power, you get 1:

**EXAMPLE**

» $m^2 \div m^2 = m^{2-2} = m^0 = 0$.

» $4x^3y^4z^7 \div 2x^3y^3z^7 = 2x^{3-3}y^{4-3}z^{7-7} = 2x^0y^1z^0 = 2y$. Both $x$ and $z$ end up with exponents of 0, so those factors become 1. Neither $x$ nor $z$ may be equal to 0.

# Taking on the Negativity of Exponents

Negative exponents are a neat little creation. They mean something very specific and have to be handled with care, but they are oh, so convenient to have.

You can use a negative exponent to write a fraction without writing a fraction! Using negative exponents is a way to combine expressions with the same base, whether the different factors are in the numerator or denominator. It's a way to change division problems into multiplication problems.

*Negative exponents* are a way of writing powers of fractions or decimals without using the fraction or decimal. For example, instead of writing $\left(\frac{1}{10}\right)^{14}$, you can write $10^{-14}$.

**ALGEBRA RULES**

The reciprocal of $x^a$ is $\frac{1}{x^a}$, which can be written as $x^{-a}$. The variable $x$ is any real number except 0, and $a$ is any real number. Also, to get rid of the negative exponent, you write:

$$x^{-a} = \frac{1}{x^a}$$

Here are two examples of changing numbers with negative exponents to fractions with positive exponents:

» $2^{-3} = \dfrac{1}{2^3} = \dfrac{1}{8}$. The reciprocal of $2^3$ is $\dfrac{1}{2^3} = 2^{-3}$.

» $6^{-1} = \dfrac{1}{6}$. The reciprocal of 6 is $\dfrac{1}{6} = 6^{-1}$.

But what if you start out with a negative exponent in the denominator? What happens then? Look at the fraction $\dfrac{1}{3^{-4}}$. If you write the denominator as a fraction, you get $\dfrac{1}{\frac{1}{3^4}}$. Then, changing the *complex fraction* (a fraction with a fraction in it) to a division problem: $\dfrac{1}{\frac{1}{3^4}} = 1 \div \dfrac{1}{3^4} = 1 \cdot \dfrac{3^4}{1} = 3^4$. So, to simplify a fraction with a negative exponent in the denominator, you can do a switcheroo: $\dfrac{1}{3^{-4}} = 3^4$.

Here are two examples of simplifying the fractions by getting rid of the negative exponents:

EXAMPLE

» $\dfrac{x^2 y^3}{3z^{-4}} = \dfrac{x^2 y^3 z^4}{3}$

» $\dfrac{4a^3 b^5 c^6 d}{a^{-1} b^{-2}} = 4a^3 a^1 b^5 b^2 c^6 d = 4a^4 b^7 c^6 d$

# Putting Powers to Work

Because exponents are symbols for repeated multiplication, one way to write $(x^3)^6$ is $x^3 \cdot x^3 \cdot x^3 \cdot x^3 \cdot x^3 \cdot x^3$. Using the multiplication rule, where you just add all the exponents together, you get $x^{3+3+3+3+3+3} = x^{18}$.

ALGEBRA
RULES

To raise a power to a power, use this formula: $(x^n)^m = x^{n \cdot m}$. In other words, when the whole expression, $x^n$, is raised to the $m$th power, the new power of $x$ is determined by multiplying $n$ and $m$ together.

Here are a few examples of simplifying using the rule for raising a power to a power:

» $\left(6^{-3}\right)^4 = 6^{(-3)(4)} = 6^{-12} = \dfrac{1}{6^{12}}$: You first multiply the exponents; then rewrite the product to create a positive exponent.

» $(x^{-2})^{-3} = x^{(-2)(-3)} = x^6$.

» $(3x^2y^3)^2 = 3^2 x^{(2)(2)} y^{(3)(2)} = 9x^4y^6$: Each factor in the parentheses is raised to the power outside the parentheses.

# Circling around Square Roots

When you do square roots, the symbol for that operation is a radical, $\sqrt{\ }$. A cube root has a small 3 in front of the radical; a fourth root has a small 4, and so on.

The radical is a non-binary operation (involving just one number) that asks you, "What number times itself gives you this number under the radical?" Another way of saying this is if $\sqrt{a} = b$, then $b^2 = a$.

Finding square roots is a relatively common operation in algebra, but working with and combining the roots isn't always so clear.

Expressions with radicals can be multiplied or divided as long as the root power *or* the value under the radical is the same. Expressions with radicals cannot be added or subtracted unless *both* the root power *and* the value under the radical are the same.

Here are some examples of simplifying the radical expressions when possible:

» $\sqrt{2} \cdot \sqrt{3} = \sqrt{6}$: These *can* be combined because it's multiplication, and the root power is the same.

» $\sqrt{8} \div \sqrt{4} = \sqrt{2}$: These *can* be combined because it's division, and the root power is the same.

» $\sqrt{2} + \sqrt{3}$: These *cannot* be combined because it's addition, and the value under the radical is not the same.

» $4\sqrt{3} + 2\sqrt{3} = 6\sqrt{3}$: These *can* be combined because the root power and the numbers under the radical are the same.

Here are the rules for adding, subtracting, multiplying, and dividing radical expressions. Assume that $a$ and $b$ are positive values.

>> $m\sqrt{a} \pm n\sqrt{a} = (m \pm n)\sqrt{a}$: Addition and subtraction can be performed if the root power and the value under the radical are the same.

>> $\sqrt{a}\sqrt{a} = \sqrt{a^2} = |a|$: The number $a$ can't be negative, so the absolute value insures a positive result.

>> $\sqrt{a}\sqrt{b} = \sqrt{ab}$: Multiplication and division can be performed if the root powers are the same.

>> $\dfrac{\sqrt{a}}{\sqrt{b}} = \sqrt{\dfrac{a}{b}}$.

When changing from radical form to fractional exponents:

>> $\sqrt[n]{a} = a^{\frac{1}{n}}$: The $n$th root of $a$ can be written as a fractional exponent with $a$ raised to the reciprocal of that power.

>> $\sqrt[n]{a^m} = a^{\frac{m}{n}}$: When the $n$th root of $a^m$ is taken, it's raised to the $\frac{1}{n}$th power. Using the "powers of powers" rule, the $m$ and the $\frac{1}{n}$ are multiplied together.

This rule involving changing radicals to fractional exponents allows you to simplify the following expressions.

Here are some examples of simplifying each expression, combining like factors:

>> $6x^2 \cdot \sqrt[3]{x} = 6x^2 \cdot x^{\frac{1}{3}} = 6x^{2+\frac{1}{3}} = 6x^{\frac{7}{3}}$.

>> $3\sqrt{x} \cdot \sqrt[4]{x^3} \cdot x = 3x^{\frac{1}{2}} \cdot x^{\frac{3}{4}} \cdot x^1 = 3x^{\frac{1}{2}+\frac{3}{4}+1} = 3x^{\frac{9}{4}}$: Leave the exponent as $\frac{9}{4}$. Don't write the exponent as a mixed number.

>> $4\sqrt{x} \cdot \sqrt[3]{a} = 4x^{\frac{1}{2}}a^{\frac{1}{3}}$: The exponents can't really be combined, because the bases are not the same.

IN THIS CHAPTER

» Applying the order of operations

» Considering the operations with constants and variables

» Distributing over two terms or many

# Chapter 3

# Ordering and Distributing: The Business of Algebra

The *order of operations* is a biggie that you use frequently when working in algebra. It tells you what to do first, next, and last in a problem, whether terms are in grouping symbols or raised to a power.

And then, after paying attention to the order of operations, you find that algebra is full of converse actions. First, you're asked to factor, and then to *distribute* or "unfactor." Distributing is a way of changing a product into a sum or difference, which allows you to combine terms and do other exciting algebraic processes.

## Taking Orders for Operations

In algebra, the order used in expressions with multiple operations depends on which mathematical operations are performed. If you're doing only addition or you're doing only multiplication, you can use any order you want. But as soon as you mix things up with addition and multiplication in the same expression, you have to pay close attention to the correct order. Mathematicians designed rules so that anyone reading a mathematical expression would do it the same way as everyone else and get the same *correct* answer. In the case of multiple signs and operations, working out the problems needs to be done in a specified *order*, from the first to the last. This is the *order of operations*.

**ALGEBRA RULES**

According to the order of operations, work out the operations and signs in the following order:

1. **Powers and roots**
2. **Multiplication and division**
3. **Addition and subtraction**

If you have more than two operations of the same level, do them in order from left to right, following the order of operations. Also, if you have any grouping symbols, perform the operations inside the grouping symbols before using the result in the order of operations.

**EXAMPLE**

Simplify the following expression using the order of operations: $6^2 - 5 \cdot 4 + 2\sqrt{16} + 24 \div 6 - 5$.

Perform the power and root first:

$$6^2 - 5 \cdot 4 + 2\sqrt{16} + 24 \div 6 - 5 = 36 - 5 \cdot 4 + 2 \cdot 4 + 24 \div 6 - 5$$

A multiplication symbol is introduced when the radical is removed — to show that the 2 multiplies the result. Two multiplications and a division are performed to get

$$36 - 20 + 8 + 4 - 5$$

Now subtract and add:

$$16 + 8 + 4 - 5 = 23$$

When you have several operations of the same "level," you perform them moving from left to right through the expression.

**EXAMPLE**

Simplify the expression: $[8 \div (5 - 3)] \cdot 5$.

$$[8 \div (5 - 3)] \cdot 5 =$$

You have to perform the operations inside the parentheses and then the bracket before multiplying by 5:

$$[8 \div 2] \cdot 5 = 4 \cdot 5 = 20$$

# Dealing with Distributing

Distributing items is the act of spreading them out equally. Algebraic distribution means to multiply each of the terms within the parentheses by another term that is outside the parentheses. Each term gets multiplied by the same amount.

**ALGEBRA RULES**

To distribute a term over several other terms, multiply each of the other terms by the first. Distribution is multiplying each individual term in a grouped series of terms by a value outside the grouping.

$$a(b + c + d + e + \ldots) = ab + ac + ad + ae + \ldots$$

The addition signs could just as well be subtraction; and $a$ is any real number: positive, negative, integer, or fraction.

Distribute the number 2 over the terms $4x + 3y - 6$.

**EXAMPLE**

1. **Multiply each term by the number(s) and/or variable(s) outside the parentheses.**

$$2(4x + 3y - 6)$$
$$2(4x) + 2(3y) - 2(6)$$

2. **Perform the multiplication operation in each term.**

$$8x + 6y - 12$$

When a number is distributed over terms within parentheses, you multiply each term by that number. And then there are the signs: Positive (+) and negative (−) signs are simple to distribute, but distributing a negative sign can create errors.

**ALGEBRA RULES**

When distributing a negative sign, each term has a change of sign: from negative to positive or from positive to negative.

Distribute $-1$ over the terms in the parentheses: $-(4x+2y-3z+7)$ is the same as multiplying through by $-1$:

$$-1(4x+2y-3z+7) =$$
$$-1(4x)-1(2y)-1(-3z)-1(7) =$$
$$-4x-2y+3z-7$$

Each term was changed to a term with the opposite sign.

Simplify the expression by distributing and combining like terms: $4x(x-2)-(5x+3)$. Treat the subtraction symbol as a distribution of $-1$ over the terms in the parentheses.

Distribute the $4x$ over the $x$ and the $-2$ by multiplying both terms by $4x$:

$$4x(x-2) = 4x(x)-4x(2)$$

Distribute the negative sign over the $5x$ and the $3$ by changing the sign of each term. Be careful — you can easily make a mistake if you stop after only changing the $5x$.

$$-(5x+3) = -(+5x)-(+3)$$

Multiply and combine the like terms:

$$4x(x)-4x(2)-(+5x)-(+3) =$$
$$4x^2-8x-5x-3 = 4x^2-13x-3$$

# Making Numbers and Variables Cooperate

Distributing variables over the terms in an algebraic expression involves multiplication rules and the rules for exponents. When different variables are multiplied together, they can be written side by side without using any multiplication symbols between them. If the same variable is multiplied as part of the distribution, then the exponents are added together. Let me show you a couple of distribution problems involving factors with exponents.

Distribute the $a$ through the terms in the parentheses: $a(a^4+2a^2+3)$.

Multiply $a$ times each term:

$$a(a^4 + 2a^2 + 3) = a \cdot a^4 + a \cdot 2a^2 + a \cdot 3$$

Use the rules of exponents to simplify:

$$a^5 + 2a^3 + 3a$$

Distribute $z^4$ over the terms in the expression $2z^2 - 3z^{-2} + z^{-4} + 5z^{\frac{1}{3}}$.

Distribute the $z^4$ by multiplying it times each term:

$$z^4\left(2z^2 - 3z^{-2} + z^{-4} + 5z^{\frac{1}{3}}\right) = z^4 \cdot 2z^2 - z^4 \cdot 3z^{-2} + z^4 \cdot z^{-4} + z^4 \cdot 5z^{\frac{1}{3}}$$

Simplify by adding the exponents:

$$2z^{4+2} - 3z^{4-2} + z^{4-4} + 5z^{4+\frac{1}{3}} =$$
$$2z^6 - 3z^2 + z^0 + 5z^{\frac{13}{3}} = 2z^6 - 3z^2 + 1 + 5z^{\frac{13}{3}}$$

The exponent 0 means the value of the expression is 1. $x^0 = 1$ for any real number $x$ except 0.

You combine exponents with different signs by using the rules for adding and subtracting signed numbers. Fractional exponents are combined after finding common denominators. Exponents that are improper fractions are left in that form.

## Relating negative exponents to fractions

As the heading suggests, a base that has a negative exponent can be changed to a fraction. The base and the exponent become part of the denominator of the fraction, but the exponent loses its negative sign in the process. Then you cap it all off with a 1 in the numerator.

The formula for changing negative exponents to fractions is $a^{-n} = \frac{1}{a^n}$. (See Chapter 2 for more details on negative exponents.)

In the following example, I show you how a negative exponent leads to a fractional answer.

Distribute the $5a^{-3}b^{-2}$ over each term in the parentheses:

$$5a^{-3}b^{-2}(2ab^3 - 3a^2b^2 + 4a^4b - ab) =$$
$$5a^{-3}b^{-2}(2ab^3) - (5a^{-3}b^{-2})(3a^2b^2) + (5a^{-3}b^{-2})(4a^4b)$$
$$(5a^{-3}b^{-2})(ab)$$

Multiplying the numbers and adding the exponents:

$$10a^{-3+1}b^{-2+3} - 15a^{-3+2}b^{-2+2} + 20a^{-3+4}b^{-2+1} - 5a^{-3+1}b^{-2+1}$$

The factor of $b$ with the 0 exponent becomes 1:

$$10a^{-2}b^1 - 15a^{-1}b^0 + 20a^1b^{-1} - 5a^{-2}b^{-1}$$

This next step shows the final result without negative exponents — using the formula for changing negative exponents to fractions (see earlier in this section):

$$\frac{10b}{a^2} - \frac{15}{a} + \frac{20a}{b} - \frac{5}{a^2b}$$

## Creating powers with fractions

Exponents that are fractions work the same way as exponents that are integers. When multiplying factors with the same base, the exponents are added together. The only hitch is that the fractions must have the same denominator to be added. (The rules don't change just because the fractions are exponents.)

Distribute and simplify: $x^{\frac{1}{4}}y^{\frac{2}{3}}\left(x^{\frac{1}{2}} + x^{\frac{1}{4}}y^{\frac{1}{3}} - y^{-\frac{1}{3}}\right)$.

Multiply the factor times each term:

$$x^{\frac{1}{4}}y^{\frac{2}{3}} \cdot x^{\frac{1}{2}} + x^{\frac{1}{4}}y^{\frac{2}{3}} \cdot x^{\frac{1}{4}}y^{\frac{1}{3}} - x^{\frac{1}{4}}y^{\frac{2}{3}} \cdot y^{-\frac{1}{3}}$$

Rearrange the variables and add the exponents:

$$x^{\frac{1}{4}}x^{\frac{1}{2}}y^{\frac{2}{3}} + x^{\frac{1}{4}}x^{\frac{1}{4}}y^{\frac{2}{3}}y^{\frac{1}{3}} - x^{\frac{1}{4}}y^{\frac{2}{3}}y^{-\frac{1}{3}} = x^{\frac{1}{4}+\frac{1}{2}}y^{\frac{2}{3}} + x^{\frac{1}{4}+\frac{1}{4}}y^{\frac{2}{3}+\frac{1}{3}} - x^{\frac{1}{4}}y^{\frac{2}{3}-\frac{1}{3}}$$

Finish up by adding the fractions:

$$x^{\frac{3}{4}}y^{\frac{2}{3}} + x^1y^1 - x^{\frac{1}{4}}y^{\frac{1}{3}}$$

Simplify by distributing: $\sqrt{xy^3}\left(\sqrt{x^5y} - \sqrt{xy^7}\right)$.

Change the radical notation to fractional exponents:

$$\sqrt{xy^3}\left(\sqrt{x^5y}-\sqrt{xy^7}\right)=\left(xy^3\right)^{\frac{1}{2}}\left[\left(x^5y\right)^{\frac{1}{2}}-\left(xy^7\right)^{\frac{1}{2}}\right]$$

Raise the powers of the products inside the parentheses:

$$x^{\frac{1}{2}}y^{\frac{3}{2}}\left[x^{\frac{5}{2}}y^{\frac{1}{2}}-x^{\frac{1}{2}}y^{\frac{7}{2}}\right]$$

Distribute the outside term over each term within the parentheses:

$$x^{\frac{1}{2}}y^{\frac{3}{2}}\left(x^{\frac{5}{2}}y^{\frac{1}{2}}\right)-x^{\frac{1}{2}}y^{\frac{3}{2}}\left(x^{\frac{1}{2}}y^{\frac{7}{2}}\right)$$

Add the exponents of the variables:

$$x^{\frac{6}{2}}y^{\frac{4}{2}}-x^{\frac{2}{2}}y^{\frac{10}{2}}$$

Simplify the fractional exponents:

$$x^3y^2-x^1y^5$$

# Making Distributions over More than One Term

The preceding sections in this chapter describe how to distribute one term over several others. This section shows you how to distribute a *binomial* (a polynomial with two terms). This same procedure can be used to distribute polynomials with three or more terms.

Distributing two terms (a *binomial*) over several terms amounts to just applying the distribution process twice. Following is an example with the steps telling you how to distribute a binomial over some polynomial.

Multiply using distribution: $(x^2+1)(y-2)$.

**EXAMPLE** 1. **Break the binomial into its two terms.**

In this case, $(x^2+1)(y-2)$, break the first binomial into its two terms, $x^2$ and 1.

## 2. Distribute each term over the other factor.

Multiply the first term, $x^2$, times the second binomial, and multiply the second term, 1, times the second binomial:

$$x^2(y-2)+1(y-2)$$

## 3. Do the two distributions.

$$x^2(y-2)+1(y-2)=x^2y-2x^2+y-2$$

## 4. Simplify and combine any like terms.

In this case, nothing can be combined; none of the terms is like any other.

When distributing a polynomial (many terms) over any number of other terms, multiply each term in the first factor times each term in the second factor. When the distribution is done, combine anything that goes together to simplify.

$$(a+b+c+d+\ldots)(z++y+x+w+\ldots)=$$
$$az+ay+ax+aw+\ldots+bz+by+bx+bw+\ldots+cz+cy+$$
$$cx+cw+\ldots+dz+dy+dx+dw+\ldots$$

# Chapter **4**

# Factoring in the First and Second Degrees

I n this chapter, you discover how to change from several terms to one, compact product. The factoring patterns you see here carry over somewhat in more complicated expressions.

When factoring, you find linear expressions (such as $2xy + 4xz$) and quadratic expressions (such as $3x^2 - 12$ or $-16t^2 + 32t + 11$). These are *expressions* because they're made up of two or more terms with plus (+) or minus (−) signs between them.

Some quadratic expressions may have one variable in them, such as $2x^2 - 3x + 1$. Others may have two or more variables, such as $\pi r^2 + 2\pi rh$.

These expressions all have their place in mathematics and science. In this chapter, you see how they work for you and how to factor them.

## Making Factoring Work

Factoring is another way of saying: "Rewrite this so everything is all multiplied together." You usually start out with an expression of two or more terms and have to determine how to rewrite them so they're all multiplied together in some way or another. And, oh yes, the two expressions have to be equal!

# Facing the factoring method

Factoring is the opposite of distributing; it's "undistributing." When performing distribution, you multiply a series of terms by a common multiplier. Now, by factoring, you seek to find what a series of terms have in common and then move it, dividing the common factor or multiplier out from each term.

**ALGEBRA RULES**

An expression can be written as the product of the largest number that divides all the terms evenly times the results of the divisions: $ab + ac + ad = a(b + c + d)$.

The absolutely *proper* way to factor an expression is to write the prime factorization of each of the numbers and look for the greatest common factor (GCF), which is the largest possible divisor shared by all the terms. What's really more practical and quicker in the end is to look for the biggest factor that *you can easily recognize*. Factor it out and then see if the numbers in the parentheses need to be factored again. Repeat the division until the terms in the parentheses are relatively prime.

**EXAMPLE**

Here's how to use the repeated division method to factor the expression $450x + 540y - 486z + 216$. You see that the coefficient of each term is even, so divide each term by 2:

$$450x + 540y - 486z + 216 =$$
$$2(225x + 270y - 243z + 108)$$

The numbers in the parentheses are a mixture of odd and even, so you can't divide by 2 again. But the numbers in the parentheses are all divisible by 9. So, factoring out 9:

$$2(225x + 270y - 243z + 108) =$$
$$2[9(25x + 30y - 27z + 12)]$$

Now multiply the 2 and 9 together to get

$$450x + 540y - 486z + 216 =$$
$$18(25x + 30y - 27z + 12)$$

You could've divided 18 into each term in the first place, but not many people know the multiplication table of 18. (It's a stretch even for me.) Looking at the terms in the parentheses, there's no single factor that divides *all* the coefficients evenly. The four coefficients are *relatively prime* (there is no one factor that they all share except the number 1), so you're finished with the factoring.

## Factoring out numbers and variables

Variables represent values; variables with exponents represent the powers of those same values. For that reason, variables as well as numbers can be factored out of the terms in an expression, and in this section I tell you how.

**ALGEBRA RULES**

When factoring out powers of a variable, the smallest power that appears in any one term is the most that can be factored out. For example, in an expression such as $a^4b + a^3c + a^2d + a^3e^4$, the smallest power of $a$ that appears in any term is the second power, $a^2$. So you can factor out $a^2$ from all the terms because $a^2$ is the GCF. You can't factor anything else out of each term: $a^4b + a^3c + a^2d + a^3e^4 = a^2(a^2b + a^1c + d + a^1e^4)$.

**EXAMPLE**

Factor the GCF out of the expression $x^2y^3 + x^3y^2z^4 + x^4yz$.

$$x^2y^3 + x^3y^2z^4 + x^4yz = x^2y(y^2 + x^1y^1z^4 + x^2z)$$

The GCF is $x^2y$.

The real test of the factoring process is combining numbers and variables, finding the GCF, and factoring successfully.

**EXAMPLE**

Factor $12x^2y^3z + 18x^3y^2z^2 - 24xy^4z^3$.

Each term has a coefficient that's divisible by 2, 3, and 6. You select 6 as the largest of those common factors.

Each term has a factor of $x$. The powers on $x$ are 2, 3, and 1. You have to select the *smallest* exponent when looking for the GCF. That's 1, so the common factor is just $x$.

Each term has a factor of $y$. The exponents are 3, 2, and 4. The smallest exponent is 2, so the common factor is $y^2$.

Each term has a factor of $z$, and the exponents are 1, 2, and 3. The number 1 is smallest, so you can pull out a $z$ from each term.

Put all the factors together, and you get that the GCF is $6xy^2z$. So,

$$12x^2y^3z + 18x^3y^2z^2 - 24xy^4z^3 =$$
$$6xy^2z(2x^1y^1 + 3x^2z^1 - 4y^2z^2)$$

Doing a quick check, multiply through by the GCF in your head to be sure that the products match the original expression. Then do a sweep to be sure that there isn't a common factor among the terms within the parentheses.

**EXAMPLE**

Factor $-4ab - 8a^2b - 12ab^2$.

The coefficient of each term in the expression is negative; dividing out the negative from all the terms in the parentheses makes them positive.

$$-4ab - 8a^2b - 12ab^2 = -4ab(1 + 2a^1 + 3b^1)$$

**REMEMBER**

When factoring out a negative factor, be sure to change the signs of each of the terms in the parentheses.

# Getting at the Basic Quadratic Expression

The word *quadratic* refers to an expression that contains a power of 2. Yes, the prefix *quad* means four, but the origin of the word *quadratic* comes from the two dimensions of a square (which is four-sided).

**ALGEBRA RULES**

The quadratic, or second-degree, expression in $x$ has the $x$ variable that is squared, and no $x$ terms with powers higher than 2. The coefficient on the squared variable is not equal to 0. The standard quadratic form is $ax^2 + bx + c$.

By convention, the terms are usually written with the second-degree term first, the first-degree term next, and the number last. After you find the variable that's squared, write the rest of the expression in decreasing powers of that variable.

**EXAMPLE**

Rewrite $aby + cdy^2 + ef$ using the standard convention involving order. This is a second-degree expression in $y$.

Written in the standard form for quadratics, $ax^2 + bx + c$, where the second-degree term comes first, it looks like

$$cdy^2 + aby + ef$$

# Following Up on FOIL and unFOIL

FOIL is an acronym used to help you multiply two binomials together by distributing. The $F$ stands for *first*; $O$, for *outer*; $I$, for *inner*; and $L$, for *last*. This section involves undoing what is done by FOILing — you see how to factor a quadratic back into the two binomials.

Many quadratic expressions, such as $6x^2 + 7x - 3$, are the result of multiplying two *binomials* (expressions with two terms), so you can undo the multiplication by factoring them:

$$6x^2 + 7x - 3 = (2x + 3)(3x - 1)$$

The product of the binomials results in the trinomial, because, using FOIL, the product of the two first terms, the $2x$ and the $3x$, is $6x^2$. Then the product of the two outer terms, $2x$ and $-1$, is $-2x$. Add that to the product of the two inner terms, $9x$, and you get $7x$. The final term is the $-3$, which comes from the product of the two last terms.

When you look at an expression such as $2x^2 - 5x - 12$, you may think that figuring out how to factor this into the product of two binomials is an awful chore. And you may wonder whether it can even be factored that way. The nice thing in solving this particular puzzle is that there's a system to make unFOILing simple. You go through the system, and it helps you find what the answer is or even helps you determine if there isn't an answer. This can't be said about all factoring problems, but it is true of quadratics in the form $ax^2 + bx + c$. That's why quadratics are so nice to work with in algebra.

The key to unFOILing these factoring problems is being organized:

» Be sure you have an expression in the form $ax^2 + bx + c$.

» Be sure the terms are written in the order of decreasing powers.

» If needed, review a listing of prime numbers and perfect squares.

» Follow the steps.

Follow these steps to factor the quadratic $ax^2 + bx + c$, using unFOIL.

**1.** **Determine all the ways you can multiply two numbers to get $a$.**

Every number can be written as at least one product, even if it's only the number times 1. So assume that there are two numbers, $e$ and $f$, whose product is equal to $a$. These are the two numbers you want for this problem.

**2.** **Determine all the ways you can multiply two numbers together to get $c$.**

If the value of $c$ is negative, ignore the negative sign for the moment. Concentrate on what factors result in the absolute value of $c$.

Now assume that there are two numbers, $g$ and $h$, whose product is equal to $c$. Use these two numbers for this problem.

**3.** **Now look at the sign of $c$ and your lists from Steps 1 and 2.**

- *If $c$ is positive,* find a value from your Step 1 list and another from your Step 2 list such that the sum of their product and the product of the two numbers they're paired with In those steps results in $b$.

  Find $e \cdot g$ and $f \cdot h$, such that $e \cdot g + f \cdot h = b$.

- *If $c$ is negative,* find a value from your Step 1 list and another from your Step 2 list such that the difference of their product and the product of the two numbers they're paired with from those steps results in $b$.

  Find $e \cdot g$ and $f \cdot h$, such that $e \cdot g - f \cdot h = b$.

**4.** **Arrange your choices as binomials.**

The $e$ and $f$ have to be in the first positions in the binomials, and the $g$ and $h$ have to be in the last positions. They have to be arranged so the multiplications in Step 3 have the correct outer and inner products.

$(e\ h)(f\ g)$

5. **Place the signs appropriately.**

   The signs are both positive if c is positive and b is positive.

   The signs are both negative if c is positive and b is negative.

   One sign is positive and one sign is negative if c is negative; deciding where to put the negative in the factorization depends on whether b is positive or negative and how you arranged the factors.

Factor the quadratic $2x^2 - 5x - 12$ using unFOIL.

**EXAMPLE**

1. **Determine all the ways you can multiply two numbers to get *a*, which is 2 in this problem.**

   The number 2 is prime, so the only way to multiply and get 2 is $2 \cdot 1$.

2. **Determine all the ways you can multiply two numbers to get *c*, which is −12 in this problem.**

   Ignore the negative sign right now. The negative becomes important in the next step. Just concentrate on what multiplies together to give you 12.

   There are three ways to multiply two numbers together to get 12: $12 \cdot 1$, $6 \cdot 2$, and $4 \cdot 3$.

3. **Look at the sign of *c* and your lists from Steps 1 and 2.**

   Because c is negative, you find a value from Step 1 and another from Step 2 such that the difference of their product and the product of the other numbers in the pairs results in b, which is −5 in this problem.

   Use the $2 \cdot 1$ from Step 1 and the $4 \cdot 3$ from Step 2. Multiply the 1 from Step 1 times the 3 from Step 2 and then multiply the 2 from Step 1 times the 4 from Step 2.

   $(1)(3) = 3$ and $(2)(4) = 8$

   The two products are 3 and 8, whose difference is 5.

4. **Arrange the choices in binomials.**

   The following arrangement multiplies the $(1x)(2x)$ to get the $2x^2$ needed for the first product. Likewise, the 4 and 3 multiply to give you 12. The outer product is $3x$ and the inner product is $8x$, giving you the difference of $5x$.

   $(1x - 4)(2x - 3)$

5. **Place the signs to give the desired results.**

   You want the $5x$ to be negative, so you need the $8x$ product to be negative. The following arrangement accomplishes this:

   $$(1x - 4)(2x + 3) =$$
   $$2x^2 + 3x - 8x - 12 =$$
   $$2x^2 - 5x - 12$$

In the next example, all the terms are positive. The sum of the outer and inner products will be used. And there are several choices for the multipliers.

Factor: $10x^2 + 31x + 15$.

EXAMPLE

1. **Determine all the ways you can multiply two numbers to get 10.**

   The 10 can be written as $10 \cdot 1$ or $5 \cdot 2$.

2. **Determine all the ways you can multiply two numbers to get 15.**

   The 15 can be written as $15 \cdot 1$ or $5 \cdot 3$.

3. **The last term is +15, so you want the sum of the products to be 31.**

   Using the $5 \cdot 2$ and the $5 \cdot 3$, multiply $2 \cdot 3$ to get 6, and multiply $5 \cdot 5$ to get 25. The sum of 6 and 25 is 31.

4. **Arrange your choices in the binomials so the factors line up the way you want to give you the products.**

   $$(2x + 5)(5x + 3)$$

5. **Placing the signs is easy, because everything is positive.**

   $$(2x + 5)(5x + 3) =$$
   $$10x^2 + 6x + 25x + 15 =$$
   $$10x^2 + 31x + 15$$

# Making UnFOIL and the GCF Work Together

A quadratic, such as $40x^2 - 40x - 240$, can be factored using two different techniques, unFOILing and the GCF, which can be done in two different orders. One of the choices makes the problem easier. It's the order in which the factoring is done that makes one way easier and the other way harder. You just have to hope that you recognize the easier way before you get started.

If you should choose to use unFOIL, first, you have to deal with the four different ways of factoring 40 and the ten ways of factoring 240, finally giving you (after much computing) $(4x-12)(10x+20)$. And then you get to factor each of the two binomials, resulting in $40(x-3)(x+2)$. It took two types of factorization: unFOILing and taking out a GCF.

An easier way is to factor out the GCF *first*.

Factor $40x^2 - 40x - 240$ by using the GCF first.

Each term's coefficient is evenly divisible by 40. Doing the factorization:

$$40x^2 - 40x - 240 = 40(x^2 - x - 6)$$

Now, looking at the trinomial in the parentheses:

1. **Use unFOIL to factor the trinomial $x^2 - x - 6$.**

   Notice how the list of choices for factors of the coefficients is much shorter and more manageable than if you try to unFOIL before factoring out the GCF.

2. **Looking at the sign of the last term, −6, choose your products to create a difference of 1.**

   The middle term, $x$, is negative, so you want the $3x$, the product of the inner terms, to be negative. Finish the factoring. Then put the 40 that you factored out in the first place back into the answer.

   $$40x^2 - 40x - 240 = 40(x-3)(x+2)$$

# Getting the Best of Binomials

If a *binomial* (two-term) expression can be factored at all, it will be factored in one of four ways. First, look at the addition or subtraction sign that always separates the two terms within a binomial. Then look at the two terms. Are they squares? Are they cubes?

Here are the four ways to factor a binomial:

>> Finding the GCF

>> Factoring the difference of two perfect squares

>> Factoring the difference of two perfect cubes

>> Factoring the sum of two perfect cubes

When you have a factoring problem with two terms, you can go through the list to see which method works. You see how to divide out a GCF earlier in this chapter, so here I show you the other three methods.

## Facing up to the difference of two perfect squares

If two terms in a binomial are perfect squares and they're separated by subtraction, then the binomial can be factored. To factor one of these binomials, just find the square roots of the two terms that are perfect squares and write the factorization as the sum and difference of the square roots.

If subtraction separates two squared terms, then the product of the sum and difference of the two square roots factors the binomial: $a^2 - b^2 = (a+b)(a-b)$.

Factor $9x^2 - 16$.

The square roots of $9x^2$ and 16 are $3x$ and 4, respectively. The sum of the roots is $3x + 4$ and the difference between the roots is $3x - 4$. So, $9x^2 - 16 = (3x+4)(3x-4)$.

Factor $25z^2 - 81y^2$.

The square roots of $25z^2$ and $81y^2$ are $5z$ and $9y$, respectively. So, $25z^2 - 81y^2 = (5z+9y)(5z-9y)$.

Factor $x^4 - y^6$.

The square roots of $x^4$ and $y^6$ are $x^2$ and $y^3$, respectively. So the factorization of $x^4 - y^6 = (x^2 + y^3)(x^2 - y^3)$.

# Creating factors for the difference of perfect cubes

A *perfect cube* is the number you get when you multiply a number times itself and then multiply the answer times the first number again. A cube is the third power of a number or variable. The difference of two cubes is a binomial expression $a^3 - b^3$.

**ALGEBRA RULES**

To factor the difference of two perfect cubes, use the following pattern: $a^3 - b^3 = (a-b)(a^2 + ab + b^2)$.

Here are the results of factoring the difference of perfect cubes:

» A binomial factor $(a - b)$ made up of the two cube roots of the perfect cubes separated by a minus sign.

» A trinomial factor $(a^2 + ab + b^2)$ made up of the squares of the two cube roots from the first factor added to the product of the cube roots in the middle. ***Remember:*** A trinomial has three terms, and this one has all plus signs in it

**EXAMPLE**

Factor $64x^3 - 27y^6$.

The cube root of $64x^3$ is $4x$, and the cube root of $27y^6$ is $3y^2$. The square of $4x$ is $16x^2$, the square of $3y^2$ is $(3y^2)^2 = 9y^4$, and the product of $(4x)(3y^2)$ is $12xy^2$.

$$64x^3 - 27y^6 = (4x - 3y^2)(16x^2 + 12xy^2 + 9y^4)$$

# Finishing with the sum of perfect cubes

You have a break coming. The rule for factoring the sum of two perfect cubes is almost the same as the rule for factoring the difference between perfect cubes, which I cover in the previous section. You just have to change two little signs to make it work.

**ALGEBRA RULES**

To factor the sum of two perfect cubes, use the following pattern: $a^3 + b^3 = (a+b)(a^2 - ab + b^2)$.

Factor $1,000z^3 + 343$.

**EXAMPLE**

The cube root of $1,000z^3$ is $10z$, and the cube root of $343$ is $7$. The product of $10z$ and $7$ is $70z$. So, $1,000z^3 + 343 = (10z + 7)(100z^2 - 70z + 49)$.

# Chapter 5

# Broadening the Factoring Horizon

This chapter has some mighty helpful factoring information that doesn't belong under linear or quadratic factoring rules. Half of the factoring process is knowing how to use the rules, and the other half is recognizing when to use what rule. These skills are equally important — you need both to be successful.

## Grabbing onto Grouping

Factoring by grouping is a way of dealing with particular four-term or six-term expressions. You identify common factors in *groups* of terms — not all of them at once. Then, if grouping is to work, you find a new common factor in the newly created terms.

### Getting the groups together

The easiest way to explain grouping is to go through an example, step by step, so you see how it works.

Factor the expression $x^5 + 7x - 14 - 2x^4$ using grouping.

**EXAMPLE**   **1.** **Write your expression in decreasing powers of a variable.**

The expression becomes $x^5 - 2x^4 + 7x - 14$.

**2.** **Determine a common factor for the first two terms and then a common factor for the second two terms. Do the factorization.**

The first two terms have a common factor of $x^4$, and the second two terms have a common factor of 7.

$$x^5 - 2x^4 + 7x - 14 = x^4(x-2) + 7(x-2)$$

**3.** **Determine if the newly written version of the expression contains terms with a common factor. Do the factorization.**

The new expression has two terms, each with a factor of $(x-2)$.

$$x^4(x-2) + 7(x-2) = (x-2)(x^4+7)$$

You should always check to see if one or both of the factors can themselves be factored. In this case, both binomials are prime, so you're done.

## Grouping and unFOILing in the same package

In the next example, you see six terms — a type of expression that uses grouping to factor it. And a big surprise comes after the grouping is finished.

Grouping isn't used on the three-term quadratic. It's usually used on expressions with four or six or some other even number of terms. When grouping is employed, you divide up the expression and factor just one part at a time. The next example walks you through the steps.

**EXAMPLE**

Factor: $3x^2y - 24xy - 27y - 5x^2z + 40xz + 45z$.

You see three different variables, and the only variable with a power greater than 1 is the $x$ variable. You rewrite the expression as

$$3x^2y - 5x^2z - 24xy + 40xz - 27y + 45z$$

You see a common factor of $x^2$ in the first two terms, a common factor of $8x$ in the third and fourth terms, and a common factor of $-9$ in the last two. (By the way, I'm purposely leading you astray for a moment and not factoring out $-8x$ in the middle grouping — just to show you how to *repair* situations when needed.) Do the three factorizations next.

$$x^2(3y-5z)+8x(-3y+5z)-9(3y-5z)$$

You see that the three binomials are not the same. In order for grouping to work, you have to create a fewer number of terms — each with some factor in common. The problem here is that you didn't factor out $-8x$. Changing the $+8x$ to $-8x$, you factor $-1$ out of each term in the second binomial and basically just change each sign to its opposite:

$$=x^2(3y-5z)-8x(3y-5z)-9(3y-5z)$$

Now you see that each of the three terms has a common factor of the binomial $(3y-5z)$. Now factor that GCF out of the three terms:

$$=(3y-5z)(x^2-8x-9)$$

Finally, the trinomial itself can be factored using unFOIL. You want the difference of the cross-products to be 8, so you use the factors $1\cdot1$ and $9\cdot1$:

$$=(3y-5z)(x-9)(x+1)$$

Yes, that's my big surprise — getting to unFOIL after grouping and factoring out several greatest common factors.

# Tackling Multiple Factoring Methods

Any factoring problem is a matter of recognizing what you have so you know what method to apply. With trinomials, you can use unFOIL if the trinomial is of the form $ax^2+bx+c$. You can find the GCF of a trinomial if a common factor is available. When you have a binomial, you look for sums or differences of cubes and differences of squares. What I show you in this section is how the different methods often appear together, and what to do when the problem needs more than one method of factoring.

When factoring, determine what type of expression you have — binomial, trinomial, squares, cubes, and so on. This helps you decide what method to use. Keep going, checking inside all parentheses for more factoring opportunities, until you're done.

# Beginning with binomials

This first example starts with a binomial. You see two squared factors in among the others, and that's it, so you don't expect anything exciting to happen. Oh, foolish you. Take the GCF out, and you find the difference of perfect cubes.

**EXAMPLE**

Factor $4x^4y - 108xy$.

The GCF of the two terms is $4xy$. Factor that out of each term first:

$$4x^4y - 108xy = 4xy(x^3 - 27)$$

Now you see that the binomial in the parentheses is the difference of two perfect cubes and can be factored using the rule from earlier in this chapter:

$$4xy(x^3 - 27) = 4xy(x - 3)(x^2 + 3x + 9)$$

Even though the last factor, the trinomial, seems to be a candidate for unFOIL, you don't have to bother. When you get a trinomial from factoring cubes, it's almost always prime. The only thing that may factor them is finding a GCF.

**EXAMPLE**

Factor $16x^4y^5(81 - z^4) - 54xy^2(81 - z^4)$.

The first thing that should jump out at you is that you see a common binomial factor of $(81 - z^4)$. Then, looking closer, you see that both terms contain factors of 2 and powers of $x$ and $y$. Factoring out the GCF,

$$16x^4y^5(81 - z^4) - 54xy^2(81 - z^4) =$$
$$2xy^2(81 - z^4)(8x^3y^3 - 27)$$

Now the factored form contains two binomial factors that can be factored. The first binomial is the difference of two squares, and the second binomial is the difference of cubes. Factoring,

$$= 2xy^2(9 - z^2)(9 + z^2)(2xy - 3)(4x^2y^2 + 6xy + 9)$$

And, of course, you realize that you're not finished. That first binomial can be factored as the difference of squares. Finishing the factoring,

$$= 2xy^2(3 - z)(3 + z)(9 + z^2)(2xy - 3)(4x^2y^2 + 6xy + 9)$$

Whew!

# Finishing with binomials

In this section, I show you how you can start with four terms, and apply a form of grouping. The result is the difference of two squares.

Factor $x^2 + 8x + 16 - y^2$.

When using grouping, you usually divide the four or six terms into equal-size groups. Sometimes four terms can be separated into unequal groupings with three terms in one group and one term in the other. The way to spot these special types of factoring situations is to look for squares. Of course, you usually don't even look for unequal groupings unless other grouping methods have failed you.

This expression has four terms, but there's no good equal pairing of terms that will give you a set of useful common factors. Another option is to group unevenly. Group the first three terms together because they form a trinomial that can be factored. That leaves the last term by itself:

$$x^2 + 8x + 16 - y^2 = (x^2 + 8x + 16) - y^2$$

Now you can factor the trinomial in the parentheses using unFOIL:

$$(x+4)^2 - y^2$$

Notice that there are now two terms and that each is a perfect square.

Using the rule $a^2 - b^2 = (a+b)(a-b)$, finish this example:

$$(x+4)^2 - y^2 = [(x+4)+y][(x+4)-y]$$

There's no big advantage to dropping the parentheses inside the brackets, so leave the answer the way it is.

# Recognizing when you have a quadratic-like expression

This next example actually incorporates a factoring technique called *quadratic-like factorization*. The trinomials you factor using this method have a resemblance to the $ax^2 + bx + c$ format with one difference: Instead of exponents of 2 and 1 on the $x$ variables, you have exponents of $2n$ and $n$ — the first exponent is twice the second.

**ALGEBRA RULES**

The quadratic-like equation $ax^{2n} + bx^n + c$ may factor into the product of two binomials of the form $(dx^n + e)(fx^n + g)$.

Factor $x^4 - 104x^2 + 400$.

**EXAMPLE**

There's no GCF, but you recognize that the first exponent is twice the second, so it fits into the pattern of a quadratic-like expression. It factors as follows:

$$x^4 - 104x^2 + 400 = (x^2 - 4)(x^2 - 100)$$

There are now two factors, but each of them is the difference of perfect squares:

$$(x^2 - 4)(x^2 - 100) = (x + 2)(x - 2)(x + 10)(x - 10)$$

You're finished!

Now I'll show you one more quadratic-like factorization — this time with exponents that are both negatives and fractions!

Factor the trinomial $3z^{-\frac{2}{3}} + 5z^{-\frac{1}{3}} - 2$.

**EXAMPLE**

The trinomial has the characteristic $2n$ and $n$ format for the exponents. You may be overwhelmed a bit, though, with coming up with the two binomials needed. To get rid of the distraction of the negative fractional exponents, rewrite the expression using a different letter and the normal exponents:

$$3y^2 + 5y - 2$$

Now it may be more clear that you can factor the basic trinomial $3y^2 + 5y - 2 = (3y - 1)(y + 2)$ and then apply that pattern to the beginning trinomial:

$$3z^{-\frac{2}{3}} + 5z^{-\frac{1}{3}} - 2 = \left(3z^{-\frac{1}{3}} - 1\right)\left(z^{-\frac{1}{3}} + 2\right)$$

## Knowing When Enough Is Enough

One of my favorite scenes from the movie *The Agony and the Ecstasy*, which chronicles Michelangelo's painting of the Sistine Chapel, comes when the pope enters the Sistine Chapel, looks up at the scaffolding, dripping paint, and Michelangelo perched up near the ceiling, and yells, "When will it be done?" Michelangelo's reply: "When I'm finished!"

The pope's lament can be applied to factoring problems: "When is it done?"

Factoring is done when no more parts can be factored. If you refer to the listing of ways to factor two, three, four, or more terms, then you can check off the options, discard those that don't fit, and stop when none works. After doing one type of factoring, you should then look at the values in parentheses to see if any of them can be factored.

Factor $3x^5 - 18x^3 - 81x$.

The GCF of the terms is $3x$.

$$3x^5 - 18x^3 - 81x = 3x(x^4 - 6x^2 - 27)$$

The trinomial can be unFOILed:

$$3x(x^4 - 6x^2 - 27) = 3x(x^2 - 9)(x^2 + 3)$$

The first binomial is the difference of squares:

$$3x(x^2 - 9)(x^2 + 3) = 3x(x - 3)(x + 3)(x^2 + 3)$$

You're finished!

# Recruiting the Remainder Theorem

The *remainder theorem* is used heavily when you're dealing with polynomials of high degrees and you want to graph them or find solutions for equations involving the polynomials. I go into these processes in great detail in *Algebra II For Dummies* (Wiley). For now, I pick out just the best part (lucky you) and show you how to make use of the remainder theorem and synthetic division to help you with your factoring chores.

The remainder theorem of algebra says that when you divide a polynomial by some linear binomial, the remainder resulting from the division is the same number as you'd get if you evaluated the polynomial using the opposite of the constant in the binomial.

The remainder theorem states that the remainder, $R$, resulting from dividing $P(x) = a_n x^n + a_{n-1} x^{n-1} + a_{n-2} x^{n-2} + \ldots + a_1 x^1 + a_0$ by $x + a$, is equal to $P(-a)$.

So, if you were to divide $x^3 + x^2 - 3x + 4$ by $x + 1$, the remainder would be $P(-1) = (-1)^3 + (-1)^2 - 3(-1) + 4 = -1 + 1 + 3 + 4 = 7$. This is what the long division looks like (and why you want to avoid it here):

$$x + 1 \overline{) \begin{array}{l} \phantom{x^3 + }x^2 \phantom{+ x^3} - 3x \\ x^3 + x^3 - 3x + 4 \\ \underline{-\left(x^2 + x^2\right)} \\ \phantom{x^3 +}0 - 3x + 4 \\ \phantom{x^3 + 0}\underline{-(-3x - 3)} \\ \phantom{x^3 + 0 - 3x + }7 \end{array}}$$

What you want, in factoring polynomials, is for the remainder to be a 0. No remainder means that the factor divided evenly into the polynomial. Long division can be tedious, and even the evaluation of polynomials can be a bit messy. So, *synthetic division* comes to the rescue.

## Getting real with synthetic division

Synthetic division is a way of dividing a polynomial by a first-degree binomial without all the folderol. In this case, the folderol is all the variables — you just use coefficients and constants. To divide $P(x) = a_n x^n + a_{n-1} x^{n-1} + a_{n-2} x^{n-2} + \ldots + a_1 x^1 + a_0$ by $x + a$, you list all the coefficients, $a_i$, putting in zeros for missing terms in the decreasing powers, and then put an upside-down division sign in front of your work. You change the $a$ in the binomial to its opposite and place it in the division sign. Then you multiply, add, multiply, add, and so on until all the coefficients have been added. The last number is your remainder.

EXAMPLE

Divide $x^4 + 5x^3 - 2x^2 - 28x - 12$ by $x + 3$ using synthetic division.

Write the coefficients in a row, with a –3 in front.

$$\underline{-3|}\ \ 1\ \ \ 5\ \ -2\ \ -28\ \ -12$$

Now bring the 1 down, multiply it times –3, put the result under the 5, and add. Multiply the sum by the –3, put it under the –2 and add. Multiply the sum times the –3, put the product under the –28, and so on. Yes!

$$\begin{array}{r|rrrr} -3 & 1 & 5 & -2 & -28 & -12 \\ & & -3 & -6 & 24 & 12 \\ \hline & 1 & 2 & -8 & -4 & 0 \end{array}$$

The first four numbers along the bottom are the coefficients of the quotient, and the 0 is the remainder. When using synthetic division to help you with factoring, the 0 remainder is what you're looking for. It means that the binomial divides evenly and is a factor. The polynomial can now be written:

$$= (x+3)(x^3 + 2x^2 - 8x - 4)$$

Next, you can see if the third-degree polynomial in the parentheses can be factored. (As it turns out, the trinomial is prime. Read the next section to see how that's determined.)

**WARNING**

When rewriting a polynomial in factored form after applying synthetic division, be sure to change the sign of the number you used in the division to its opposite in the binomial.

## Making good choices for synthetic division

Synthetic division is quick, neat, and relatively painless. But even quick, neat, and painless becomes tedious when you apply it without good results. When determining what might factor a particular polynomial, you need some clues. For example, you might be wondering if $(x-1)$, $(x+4)$, $(x-3)$, or some other binomials are factors of $x^4 - x^3 - 7x^2 + x + 6$. I can tell just by looking that the binomial $(x+4)$ won't work and that the other two factors are possibilities. How can I do that?

**ALGEBRA RULES**

The *rational root theorem* says that if a *rational number* (a number that can be written as a fraction) is a solution, $r$, of the equation $a_n x^n + a_{n-1} x^{n-1} + a_{n-2} x^{n-2} + \ldots + a_1 x^1 + a_0 = 0$, then $r = \dfrac{\text{some factor of } a_0}{\text{some factor of } a_n}$.

Using the rational root theorem for my factoring, I just find these possible solutions of the equation and do the synthetic division using only these possibilities.

**EXAMPLE**

Factor $x^4 - x^3 - 7x^2 + x + 6$ using synthetic division, the rational root theorem, and the factor theorem.

The *factor theorem* says that if $x = a$ is a root of a polynomial equation, then $x - a$ is a factor of the polynomial.

First, I make a list of the possible solutions if this were an equation. All the factors of the constant, $a_0$, are ±1, ±2, ±3, and ±6.

Next, I divide each of the factors by the factors of the lead coefficient, $a_n$. I caught a break here. The lead coefficient is a 1, so the divisions are just the original numbers.

Now I use synthetic division to see if I get a remainder of 0 using any of these numbers:

$$\underline{1\rfloor}\ \begin{array}{rrrrr} 1 & -1 & -7 & 1 & 6 \\ & 1 & 0 & -7 & -6 \\ \hline 1 & 0 & -7 & -6 & 0 \end{array}$$

The number 1 is a solution, so $(x - 1)$ is a factor. Dividing again, into the result:

$$-\underline{1\rfloor}\ \begin{array}{rrrr} 1 & 0 & -7 & -6 \\ & -1 & 1 & 6 \\ \hline 1 & -1 & -6 & 0 \end{array}$$

The number −1 is a solution, so $(x + 1)$ is a factor. The numbers across the bottom are the coefficients of the trinomial factor multiplying the two binomial factors, so you can write

$$x^4 - x^3 - 7x^2 + x + 6 = (x - 1)(x + 1)(x^2 - x - 6)$$

What's even nicer is that the trinomial is easily factored, giving you an end result of

$$(x - 1)(x + 1)(x - 3)(x + 2)$$

# Factoring Rational Expressions

A rational expression consists of a fraction with algebraic terms in the numerator or denominator or both. Because the fraction line acts like a grouping symbol, you have to factor the numerator and denominator separately before doing any reducing of fractions or other operations.

Reduce the fraction using a GCF: $\dfrac{4x^2 - 36x}{x^2 - 81}$

First, you factor the numerator by pulling out the greatest common factor, $4x$. The denominator is the difference of perfect squares.

$$\frac{4x^2 - 36x}{x^2 - 81} = \frac{4x(x-9)}{(x-9)(x+9)}$$

The numerator and denominator have the common factor $(x-9)$, so that can be divided out.

$$\frac{4x\cancel{(x-9)}}{\cancel{(x-9)}(x+9)} = \frac{4x}{x+9}$$

A *cardinal sin* in algebra is to reduce rational expressions incorrectly. You can only divide out factors — multipliers — but never terms. The error I show you here is a *huge* no-no:

$$\frac{4x^2 - 36x}{x^2 - 81}$$

$$\frac{\cancel{4x^2} - \cancel{36x}}{\cancel{x^2} - \cancel{81}}$$

# Chapter 6

# Solving Linear Equations

L inear equations consist of some terms that have variables and others that are constants. A standard form of a linear equation is $ax + b = c$. What distinguishes linear equations from the rest of the pack is the fact that the variables are always raised to the first power.

In this chapter, I take you through many different types of opportunities for dealing with linear equations. Most of the principles you use with these first-degree equations are applicable to the higher-order equations, so you don't have to start from scratch later on.

## Playing by the Rules

When you're solving equations with just two terms or three terms or even more than three terms, the big question is: "What do I do first?"

Actually, as long as the equation stays balanced, you can perform any operations in any order. But you also don't want to waste your time performing operations that don't get you anywhere or even make matters worse.

The following list tells you how to solve your equations in the best order. The basic process behind solving equations is to use the *reverse* of the order of operations.

REMEMBER The order of operations (see Chapter 3) is powers or roots first, then multiplication and division, and addition and subtraction last. Grouping symbols override the order — you perform the operations inside the grouping symbols to get rid of them first.

When you're solving equations, you still usually deal with the grouping symbols first, but, for the rest of the equation, you reverse the order of operations:

1. **Do all the addition and subtraction.**

   Combine all terms that can be combined both on the same side of the equation and on opposite sides using addition and subtraction.

2. **Do all multiplication and division.**

   This step is usually the one that isolates or solves for the value of the variable or some power of the variable.

3. **Multiply exponents and find the roots.**

   Powers and roots aren't found in these linear equations — they come in quadratic and higher-powered equations. But these would come next in the reverse order of operations.

When solving linear equations, the goal is to isolate the variable you're trying to find the value of. Isolating it, or getting it all alone on one side, can take one step or many steps. And it has to be done according to the rules.

# Solving Equations with Two Terms

Linear equations contain variables raised to the first power. The easiest types of linear equations to solve are those consisting of two terms, such as those in the form: $ax = b$ or $cx + d = 0$.

Linear equations that contain just two terms are solved with multiplication, division, reciprocals, or some combinations of the operations.

# Depending on division

One of the most basic methods for solving equations is to divide each side of the equation by the same number. Many formulas and equations include a *coefficient* (multiplier) with the variable. To get rid of the coefficient and solve the equation, you divide. The following example takes you step-by-step through solving with division.

Solve for $x$ in $20x = 170$.

**EXAMPLE**

1. **Determine the coefficient of the variable and divide both sides by it.**

   Because the equation involves multiplying by 20, undo the multiplication in the equation by doing the opposite, which is division. Divide each side by 20:

   $$\frac{20x}{20} = \frac{170}{20}$$

2. **Reduce both sides of the equal sign.**

   $$\frac{\cancel{20}x}{\cancel{20}} = \frac{170}{20}$$
   $$x = 8.5$$

Now, let me show you an example with a practical application embedded in it.

You need to buy 300 doughnuts for a big meeting. How many dozen doughnuts is that?

**EXAMPLE**

Let $d$ represent the number of dozen doughnuts you need. There are 12 doughnuts in a dozen, so $12d = 300$. Twelve times the number of dozens of doughnuts you need has to equal 300.

1. **Determine the coefficient of the variable and divide both sides by it.**

   Divide each side by 12.

   $$\frac{12d}{12} = \frac{300}{12}$$

2. **Reduce both sides of the equal sign.**

   $$d = 25 \text{ dozen donuts}$$

# Making use of multiplication

The opposite operation of multiplication is division. I use division in the preceding section to solve equations where a number multiplies the variable. The reverse occurs in this section: I use multiplication where a number already *divides* the variable. The first example walks you through the steps needed.

Solve for $y$ in $\frac{y}{11} = -2$.

EXAMPLE

**1. Determine the value that divides the variable and multiply both sides by it.**

In this case, 11 is dividing the $y$, so that's what you multiply by.

$$11\left(\frac{y}{11}\right) = (-2)(11)$$

**2. Reduce on the left side and multiply on the right.**

$$\not{11}\left(\frac{y}{\not{11}}\right) = -22$$

$$y = -22$$

Next, look at an example that's applicable — a bit hairy (pardon the pun), but you read about situations like this all the time.

A wealthy woman's will dictated that her fortune be divided evenly among her nine cats. Each feline got $500,000, so what was her total fortune before it was split up? (Cats don't pay inheritance tax. Does that give you paws? Ouch.)

EXAMPLE

Let $f$ represent the amount of her fortune. Then you can write the equation:

$$\frac{f}{9} = 500,00$$

In other words, the fortune divided by 9 gave a share of $500,000.

**1. Determine the value that divides the variable and multiply both sides by it.**

In this equation, the fortune was divided. Solve the puzzle by multiplying each side by 9. The opposite of division is multiplication, so multiplication undoes what division did.

$$9\left(\frac{f}{9}\right) = 500,000 \cdot 9$$

**2.** Reduce on the left and multiply on the right.

$$\cancel{g}\left(\frac{f}{\cancel{g}}\right) = 4,500,000$$

$$f = \$4,500,000$$

Her fortune was $4.5 million! Those are nine very happy kitties. You can bet their caretakers hope they have nice, long lives.

In the next example, the variable is multiplied by 4 and divided by 5. So, you solve the problem using *both* multiplication and division.

Solve for $a$ in $\frac{4a}{5} = 12$.

**1.** Determine what is dividing the variable.

In this case, the 5 is dividing both the 4 and the variable $a$.

**2.** Multiply the values on each side of the equal sign by 5.

$$5\left(\frac{4a}{5}\right) = 12 \cdot 5$$

**3.** Reduce and simplify.

$$\cancel{5}\left(\frac{4a}{\cancel{5}}\right) = 12 \cdot 5$$

$$4a = 60$$

**4.** Determine what is multiplying the variable.

The number 4 is the coefficient and multiplies the $a$.

**5.** Divide the values on each side of the equal sign.

$$\frac{4a}{4} = \frac{60}{4}$$

**6.** Reduce and simplify.

$$\frac{\cancel{4}a}{\cancel{4}} = \frac{60}{4}$$

$$a = 15$$

A simpler way of solving this last equation is to multiply by the reciprocal of the variable's coefficient. I show you that alternative next.

## Reciprocating the invitation

The reciprocal of a number is its "flip." A more mathematical definition is that a number and its reciprocal have a product of 1.

What makes the reciprocal so important in algebra is that you can create the number 1 as a coefficient of a variable by multiplying by the reciprocal of the current coefficient. So, in a way, this process is just a special case of multiplying each side by the same number.

In the following example, I solve a problem (the last example in the preceding section) using the reciprocal instead of doing the two operations of multiplication and division.

**EXAMPLE**

Solve for $a$ in $\frac{4a}{5} = 12$.

The coefficient of the variable $a$ is the fraction $\frac{4}{5}$. The reciprocal of $\frac{4}{5}$ is $\frac{5}{4}$. So, to solve for $a$, you multiply each side of the equation by $\frac{5}{4}$:

$$\frac{\cancel{5}}{\cancel{4}} \cdot \frac{\cancel{4}a}{\cancel{5}} = \frac{\overset{3}{\cancel{12}}}{1} \cdot \frac{5}{\cancel{4}}$$
$$a = 15$$

# Taking on Three Terms

The standard form of a linear equation is $ax + b = c$. In the "Solving Equations with Two Terms" section, earlier in this chapter, you have linear equations for which the value of $b$ is 0, which gives you just $ax = c$. In this section, I introduce that extra constant value and show you how to deal with it. Also, in this section, you find equations that start out with more than one variable term, and you work toward combining and creating a new equation with just the one variable.

In general, you solve linear equations by simplifying and performing operations that give you a variable term on one side of the equal sign and a constant term on the other side. Then you can use multiplication or division to finish the problem.

# Eliminating a constant term

When you have a linear equation involving three terms, and just one of the terms contains a variable factor, you add or subtract a constant to isolate that variable term — get it by itself on one side of the equation.

Solve for $y$ in $3y - 11 = 19$.

To isolate the $y$ term, you add 11 to each side of the equation. The number 11 is chosen, because it's the opposite of $-11$, and the sum of $-11$ and 11 is 0.

$$\begin{array}{r} 3y - 11 = 19 \\ +11 \; +11 \\ \hline 3y \quad\;\; = 30 \end{array}$$

Now you have a linear equation in two terms, which is solved by dividing each side of the equation by 3:

$$\frac{\cancel{3}y}{\cancel{3}} = \frac{\overset{10}{\cancel{30}}}{\cancel{3}}$$

$$y = 10$$

# Vanquishing the extra variable term

One aim of the linear-equation solver is to get the variable term on one side of the equation and the constant term on the other side. In the preceding section, I show you how to get rid of the pesky extra constant. But what if you have more than one variable term? Can that be dealt with as easily as the constant numbers? The answer is a resounding "yes."

To reduce your linear equation to one variable term, you first perform any addition or subtraction necessary to get all the variable terms on one side of the equation.

Solve for the value of $x$ in $5x - 4 = 3x + 8$.

First, subtract $3x$ from each side of the equation. That step removes the $x$ term from the right side. Subtracting $5x - 3x$, you get $2x$, because the two terms have the same variable:

$$\begin{array}{r} 5x - 4 = 3x + 8 \\ -3x \quad\; -3x \\ \hline 2x - 4 = \qquad 8 \end{array}$$

Now the problem looks like those from the previous section. I isolate the variable term by adding 4 to each side of the equation:

$$2x - 4 = \phantom{0}8$$
$$\underline{+4 \quad +4\phantom{00}}$$
$$2x \phantom{00} = 12$$

Now the problem is finished by dividing each side of the equation by 2:

$$\frac{2x}{2} = \frac{\overset{6}{\cancel{12}}}{\cancel{2}}$$
$$x = 6$$

# Breaking Up the Groups

Linear equations don't always start out in the nice, $ax + b = c$ form. Sometimes, because of the complexity of the application, a linear equation can contain multiple variable and constant terms and lots of grouping symbols, such as in this equation:

$$3[4x + 5(x + 2)] + 6 = 1 - 2[9 - 2(x - 3)]$$

The different types of grouping symbols are used for *nested expressions* (one inside the other). The rules regarding order of operations (see Chapter 3) apply as you work toward figuring out what the variable *x* represents.

## Nesting isn't for the birds

When you have a number or variable that needs to be multiplied by every value inside parentheses, brackets, braces, or a combination of those grouping symbols, you distribute that number or variable. *Distributing* means that the number or variable next to the grouping symbol multiplies every value inside the grouping symbol. If two or more of the grouping symbols are inside one another, they're nested. Nested expressions are written within parentheses, brackets, and braces to make the intent clearer.

## Distributing first

Equations containing grouping symbols offer opportunities for making wise decisions. In some cases you need to distribute, working from the inside out, and in other cases it's wise to multiply or divide first. In general, you'll distribute first if you find more than two terms in the entire equation.

Solve for $x$ in $3[4x+5(x+2)]+6=1-2[9-2(x-4)]$.

The best way to sort through all these operations is to simplify from the inside out. You see parentheses within brackets. The binomials in the parentheses have multipliers. I'll step through this carefully to show you an organized plan of attack.

First, distribute the 5 over the binomial inside the left parentheses and the $-2$ over the binomial inside the right parentheses:

$$3[4x+5x+10]+6=1-2[9-2x+8]$$

Now combine terms within the brackets:

$$3[9x+10]+6=1-2[17-2x]$$

Distribute the 3 over the two terms in the left brackets and the $-2$ over the terms in the right brackets:

$$27x+30+6=1-34+4x$$

The constant terms on each side can be combined:

$$27x+36=-33+4x$$

Now subtract $4x$ from each side and subtract 36 from each side:

$$\begin{array}{r} 27x+36=-33+4x \\ \underline{-4x \qquad\quad -4x} \\ 23x+36=-33 \\ \underline{-36 \ -36} \\ 23x \quad\ =-69 \end{array}$$

Now, dividing each side of the equation by 23, you get that $x=-3$.

## Multiplying before distributing

In this section, I show you where it might be easier to divide through by a number rather than distribute first. My only caution is that you always divide (or multiply) each term by the same number.

This example mixes two different situations that are actually the same. The terms in the equation either have a fractional multiplier or are in a fraction themselves. The point of the example is to show when multiplying each term by the same number first is preferable to distributing first.

**EXAMPLE**

Solve for $x$ in the following equation:

$$\frac{3(x-2)}{4} + \frac{1}{2}(5x+2) = \frac{14x+12}{8} + 7$$

At first glance, the equation looks a bit forbidding. But quick action — in the form of multiplying each term by 8 — takes care of all the fractions. You're left with rather large numbers, but that's still nicer than fractions with different denominators. I choose to multiply by 8, because that's the least common denominator of each term (even the last term). Each of the four terms is multiplied by 8:

$$\frac{^2\cancel{8}}{1}\left[\frac{3(x-2)}{\cancel{4}}\right] + \frac{^4\cancel{8}}{1}\left[\frac{1}{\cancel{2}}(5x+2)\right] = \frac{\cancel{8}}{1}\left[\frac{14x+12}{\cancel{8}}\right] + 8(7)$$

$$6(x-2) + 4(5x+2) = (14x+12) + 56$$

Do the multiplication and distributing in steps to avoid errors:

$$6x - 12 + 20x + 8 = 14x + 12 + 56$$

The two variable terms on the left and the two constant terms on the left can be combined. Likewise, combine the two constant terms on the right:

$$26x - 4 = 14x + 68$$

Now subtract $14x$ from each side and add 4 to each side:

$$
\begin{array}{rcl}
-26x - 4 & = & 14x + 68 \\
\underline{-14x} & & \underline{-14x} \\
-12x - 4 & = & 68 \\
\underline{+4} & & \underline{+4} \\
-12x & = & 72
\end{array}
$$

Dividing each side of the equation by 12, you see that $x = 6$.

# Focusing on Fractions

Fractions appear frequently in algebraic equations. In the "Multiplying before distributing" section, earlier in this chapter, I show you how to remove the fractions from an equation when you have the right situation. In this section, I show you how to leave in the fraction, take advantage of the fractional setup, and use it to your advantage.

## Promoting proportions

A *proportion* is an equation. It consist of two ratios (fractions) set equal to one another. When you write $\frac{6}{12} = \frac{1}{2}$, you're writing a proportion. Before I show you how proportions are solved in algebra problems, I have some properties to share.

**ALGEBRA RULES**

Given the proportion $\frac{a}{b} = \frac{c}{d}$:

» The cross-products are equal: $ad = bc$.

» The reciprocals are equal: $\frac{b}{a} = \frac{d}{c}$.

» You can reduce the fractions vertically, as usual: $\frac{\cancel{e} \cdot f}{\cancel{e} \cdot g} = \frac{c}{d}$.

» You can reduce horizontally, across the equal sign:

$$\frac{\cancel{e} \cdot f}{b} = \frac{\cancel{e} \cdot g}{d} \text{ or } \frac{a}{\cancel{e} \cdot f} = \frac{c}{\cancel{e} \cdot g}.$$

Now I'll use some of the properties of proportions to solve equations.

**EXAMPLE**

Solve for $x$: $\frac{3x-5}{x+3} = \frac{24}{15}$.

Before cross-multiplying, reduce the fraction on the right by dividing the numerator and denominator by 3:

$$\frac{3x-5}{x+3} = \frac{\overset{8}{\cancel{24}}}{\underset{5}{\cancel{15}}} = \frac{8}{5}$$

Now, using the cross-multiplying rule:

$$(3x-5)\cdot 5 = (x+3)\cdot 8$$
$$15x - 25 = 8x + 24$$

Subtract 8x from each side, and add 25 to each side:

$$15x - 25 = 8x + 24$$
$$\underline{-8x \qquad -8x}$$
$$-7x - 25 = \qquad 24$$
$$\underline{+25 \qquad +25}$$
$$7x \quad = \quad 49$$

Finally, divide each side by 7, and you get $x = 7$.

WARNING

When reducing proportions, you can divide vertically or horizontally, but you can't reduce the fractions diagonally. The diagonal reductions are done when you're multiplying fractions and you have a multiplication symbol between them, not an equal sign between them.

## Taking advantage of proportions

Proportions are very nice to work with because of their unique properties of reducing and changing into non-fractional equations. Many equations involving fractions must be dealt with in that fractional form, but other equations are easily changed into proportions. When possible, you want to take advantage of the situations where transformations can be done.

EXAMPLE

Solve the following equation for $x$:

$$\frac{x+2}{3} - \frac{5x+1}{6} = \frac{3x-1}{2} + \frac{x-9}{8}$$

You could solve the problem by multiplying each fraction by the least common factor of all the fractions: 24.

Another option is to find a common denominator for the two fractions on the left and subtract them, and then find a common denominator for the two fractions on the right and add them. Your result is a proportion:

$$\frac{2(x+2)}{2 \cdot 3} - \frac{5x+1}{6} = \frac{4(3x-1)}{4 \cdot 2} + \frac{x-9}{8}$$
$$\frac{2(x+2)-(5x+1)}{6} = \frac{4(3x-1)+(x-9)}{8}$$
$$\frac{2x+4-5x-1}{6} = \frac{12x-4+x-9}{8}$$
$$\frac{-3x+3}{6} = \frac{13x-13}{8}$$

The proportion can be reduced by dividing by 2 horizontally:

$$\frac{-3x+3}{\cancel{6}_{\,3}} = \frac{13x-13}{\cancel{8}_{\,4}}$$

Now cross-multiply and simplify the products:

$$(-3x+3)\cdot 4 = 3\cdot(13x-13)$$
$$-12x+12 = 39x-39$$

Add 12x to each side, and then add 39 to each side:

$$\begin{array}{r} -12x+12 = 39x-39 \\ +12x \qquad +12x \\ \hline 12 = 51x-39 \\ +39 \qquad +39 \\ \hline 51 = 51x \end{array}$$

The last step consists of just dividing each side by 51 to get $1 = x$.

# Changing Formulas by Solving for Variables

A formula is an equation that represents a relationship between some structures, quantities, or other entities. It's a rule that uses mathematical computations and can be counted on to be accurate each time you use it (when applied correctly).

Here are some of the more commonly used formulas that contain only variables raised to the first power:

» $A = \frac{1}{2}bh$: The area of triangle involves base and height.

» $I = Prt$: The interest earned uses principal, rate, and time.

» $C = 2\pi r$: Circumference is twice π times the radius.

» $°F = 32° + \frac{9}{5}°C$: Degrees Fahrenheit can be expressed in terms of degrees Celsius.

» $P = R - C$: Profit is based on revenue and cost.

When you use a formula to find the indicated variable (the one on the left of the equal sign), then you just put the numbers in, and out pops the answer. Sometimes, though, you're looking for one of the other variables in the equation and end up solving for that variable over and over.

EXAMPLE

For example, let's say that you're planning a circular rose garden in your backyard. You find edging on sale and can buy a 20-foot roll of edging, a 36-foot roll, a 40-foot roll, or a 48-foot roll. You're going to use every bit of the edging and let the length of the roll dictate how large the garden will be. If you want to know the radius of the garden based on the length of the roll of edging, you use the formula for circumference and solve the following four equations:

$$20 = 2\pi r \qquad 36 = 2\pi r \qquad 40 = 2\pi r \qquad 48 = 2\pi r$$

Another alternative to solving four different equations is to solve for $r$ in the formula and then put the different roll sizes into the new formula. Starting with $C = 2\pi r$, you divide each side of the equation by $2\pi$, giving you:

$$r = \frac{C}{2\pi}$$

The computations are much easier if you just divide the length of the roll by $2\pi$.

Following is an example of solving for one of the variables in an equation. I won't try to come up with any more gardening or other clever scenarios.

EXAMPLE

Solve for $w$ in the formula for the perimeter of a rectangle: $P = 2(l + w)$.

First, divide each side of the equation by 2 (instead of distributing the 2 through the terms in the binomial):

$$\frac{P}{2} = l + w$$

Now subtract $l$ from each side. You can write the two terms as a single fraction if you want:

$$\frac{P}{2} = l + w \text{ or } \frac{P - 2l}{2} = w$$

# Chapter 7

# Tackling Second-Degree Quadratic Equations

A quadratic equation is a quadratic expression (a grouping of terms whose highest variable power is 2) with an equal sign attached. As with linear equations, specific methods or processes, given in detail in this chapter, are employed to successfully solve quadratic equations. The most commonly used technique for solving these equations is factoring, but there's also a quick and dirty rule for one of the special types of quadratic equations.

Quadratic equations are important to algebra and many other sciences. Some quadratic equations say that what goes up must come down. Other equations describe the paths that planets and comets take. In all, quadratic equations are fascinating — and just dandy to work with.

## Recognizing Quadratic Equations

A quadratic equation contains a variable term with an exponent of 2 and no variable term with a higher power.

A quadratic equation has a general form that looks like this: $ax^2 + bx + c = 0$. The constants $a$, $b$, and $c$ in the equation are real numbers, and $a$ cannot be equal to 0. (If $a$ were 0, you wouldn't have a quadratic equation anymore.)

Here are some examples of quadratic equations and their solutions:

» **$4x^2 + 5x - 6 = 0$:** In this equation, none of the coefficients is 0. The two solutions are $x = -2$ and $x = \frac{3}{4}$.

» **$2x^2 - 18 = 0$:** In this equation, the $b$ is equal to 0. The solutions are $x = 3$ and $x = -3$.

» **$x^2 + 3x = 0$:** In this equation, the $c$ is equal to 0. The solutions are $x = 0$ and $x = -3$.

» **$x^2 = 0$:** In this equation, both $b$ and $c$ are equal to 0. The equation has only one solution, $x = 0$.

A special feature of quadratic equations is that they can, and often do, have two completely different answers. As you see in the preceding examples, three of the equations have different solutions. The last equation has just one solution, but, technically, you count that solution twice, calling it a *double root.* Some quadratic equations have no solutions if you're only considering real numbers, but get real! I stick to real solutions for now.

# Finding Solutions for Quadratic Equations

The general quadratic equation has the form $ax^2 + bx + c = 0$, and $b$ or $c$ or both of them can be equal to 0. This section shows you how nice it is — and how easy it is to solve equations — when $b$ is equal to 0.

The following is the rule for some special quadratic equations — the ones where $b = 0$. They start out looking like $ax^2 + c = 0$, but the $c$ is usually negative, giving you $ax^2 - c = 0$, and the equation is rewritten as $ax^2 = c$.

If $x^2 = k$, then $x = \pm\sqrt{k}$ or if $ax^2 = c$, then $x = \pm\sqrt{\frac{c}{a}}$. If the square of a variable is equal to the number $k$, then the variable is equal to either a positive or negative number — both the positive and negative roots of $k$.

The following examples show you how to use this square-root rule on quadratic equations where $b = 0$.

**EXAMPLE**

Solve for $x$ in $x^2 = 49$.

Using the square-root rule, $x = \pm\sqrt{49} = \pm7$. Checking, $(7)^2 = 49$ and $(-7)^2 = 49$.

**EXAMPLE**

Solve for $m$ in $3m^2 + 4 = 52$.

This equation isn't quite ready for the square-root rule. Add $-4$ to each side:

$$3m^2 = 48$$

Now divide each side by 3:

$$m^2 = 16$$

So $m = \pm\sqrt{16} = \pm4$.

Solve for $q$ in $(q + 3)^2 = 25$.

**EXAMPLE**

In this case, you end up with two completely different answers, not one number and its opposite. Use the square-root rule, first, to get $q + 3 = \pm\sqrt{25} = \pm5$.

Now you have two different linear equations to solve:

$$q + 3 = +5 \qquad q + 3 = -5$$

Subtracting 3 from each side of each equation, the two answers are

$$q = 2 \qquad q = -8$$

# Applying Factorizations

This section is where running through all the factoring methods can really pay off. In most quadratic equations, factoring is used rather than the square-root rule method covered in the preceding section. The square-root rule is used only when $b = 0$ in the quadratic equation $ax^2 + bx + c = 0$. Factoring is used when $c = 0$ or when neither $b$ nor $c$ is 0.

A very important property used along with the factoring to solve these equations is the multiplication property of zero. This is a very straightforward rule — and it even makes sense. Use the greatest common factor (GCF) and the multiplication property of zero when solving quadratic equations that aren't in the form for the square-root rule.

## Zeroing in on the multiplication property of zero

Before you get into factoring quadratics for solutions, you need to know about the multiplication property of zero. By itself, 0 is nothing. Put it as the result of a multiplication problem, and you really have something: the *multiplication property of zero.*

ALGEBRA
RULES

The *multiplication property of zero* (MPZ) states that if $pq = 0$, then either $p = 0$ or $q = 0$. One of them must be equal to 0 (or both could be 0).

This may seem obvious, but think about it. A product of 0 leads to one conclusion: One of the multipliers must be 0. No other means of arriving at a 0 product exists. Why is this such a big deal? Let me show you a few equations and how the MPZ works.

EXAMPLE

Find the value of $x$ if $3x = 0$.

$x = 0$ because 3 can't be 0. Using the MPZ, if the one factor isn't 0, then the other must be 0.

EXAMPLE

Find the value of $x$ and $y$ if $xy = 0$.

You have two possibilities in this equation. If $x = 0$, then $y$ can be any number, even 0. If $x \neq 0$, then $y$ must be 0, according to the MPZ.

EXAMPLE

Solve for $x$ in $x(x - 5) = 0$.

Again, you have two possibilities. If $x = 0$, then the product of $0(-5) = 0$. The other choice is when $x = 5$. Then you have $5(0) = 0$.

# Solving quadratics by factoring and applying the multiplication property of zero

Factoring is relatively simple when there are only two terms and they have a common factor. This is true in quadratic equations of the form $ax^2 + bx = 0$ (where $c = 0$). The two terms left have the common factor of $x$, at least. You find the GCF and factor that out, and then use the MPZ to solve the equation.

The following example makes use of the fact that the constant term is 0, and there's a common factor of at least an $x$ in the two terms.

**EXAMPLE**

Solve for $x$ in $6x^2 + 18x = 0$.

The GCF of the two terms is $6x$, so write the left side in factored form:

$$6x(x + 3) = 0$$

Use the MPZ to say that $6x = 0$ or $x + 3 = 0$, which gives you the two solutions $x = 0$ and $x = -3$.

Technically, I could have written three different equations from the factored form:

$$6 = 0 \qquad x = 0 \qquad x + 3 = 0$$

The first equation, $6 = 0$, makes no sense — it's an impossible statement. So you either ignore setting the constants equal to 0 or combine them with the factored-out variable, where they'll do no harm.

**WARNING**

Missing the $x = 0$, a full half of the solution, is an amazingly frequent occurrence. You don't notice the lonely little $x$ in the front of the parentheses and forget that it gives you one of the two answers. Be careful.

# Solving Three-Term Quadratics

In the two previous sections, either $b$ or $c$ has been equal to 0 in the quadratic equation $ax^2 + bx + c = 0$. Now I won't let anyone skip out. In this section, each of the letters, $a$, $b$, and $c$ is a number that is not 0.

To solve a quadratic equation, moving everything to one side with 0 on the other side of the equal sign is the most efficient method. Factor the equation if possible, and use the MPZ after you factor. If there aren't three terms in the equation, then refer to the previous sections.

In the following example, I list the steps you use for solving a quadratic trinomial by factoring.

Solve for $x$ in $x^2 - 3x = 28$. Follow these steps:

**1.  Move all the terms to one side. Get 0 alone on the right side.**

In this case, you can subtract 28 from each side:

$$x^2 - 3x - 28 = 0$$

**2.  Determine all the ways you can multiply two numbers to get $a$.**

In $x^2 - 3x - 28 = 0$, $a = 1$, which can only be 1 times itself.

**3.  Determine all the ways you can multiply two numbers to get $c$ (ignore the sign for now).**

Twenty-eight can be $1 \cdot 28$, $2 \cdot 14$, or $4 \cdot 7$.

**4.  Factor.**

If $c$ is positive, find an operation from your Step 2 list and an operation from your Step 3 list that match so that the sum of their cross-products is the same as $b$.

If $c$ is negative, find an operation from your Step 2 list and an operation from your Step 3 list that match so that the difference of their cross-products is the same as $b$.

In this problem, $c$ is negative, and the difference of 4 and 7 is 3. Factoring, you get $(x-7)(x+4) = 0$.

5. **Use the MPZ.**

Either $x - 7 = 0$ or $x + 4 = 0$; now try solving for $x$ by getting $x$ alone to one side of the equal sign.

- $x - 7 + 7 = 0 + 7$ gives you that $x = 7$.

- $x + 4 - 4 = 0 - 4$ gives you that $x = -4$.

So the two solutions are $x = 7$ and $x = -4$.

6. **Check your answer.**

If $x = 7$, then $(7)^2 - 3(7) = 49 - 21 = 28$.

If $x = -4$, then $(-4)^2 - 3(-4) = 16 + 12 = 28$.

They both check.

Factoring to solve quadratics sounds pretty simple on the surface. But factoring *trinomial equations* — those with three terms — can be a bit less simple. If a quadratic with three terms can be factored, then the product of two binomials is that trinomial. If the quadratic equation with three terms can't be factored, then use the quadratic formula (see "Calling on the Quadratic Formula" later in this chapter).

REMEMBER

The product of the two binomials $(ax + b)(cx + d)$ is equal to the trinomial $acx^2 + (ad + bc)x + bd$. This is a fancy way of showing what you get from using FOIL when multiplying the two binomials together.

Now, on to using unFOIL. If you need more of a review of FOIL and unFOIL, check out Chapter 4.

The following examples all show how factoring and the MPZ allow you to find the solutions of a quadratic equation with all three terms showing.

EXAMPLE

Solve for $x$ in $x^2 - 5x - 6 = 0$.

1. **Check whether the equation is in standard form.**

The equation is in standard form, so you can proceed.

2. **Determine all the ways you can multiply to get $a$.**

$a = 1$, which can only be 1 times itself. If there are two binomials that the left side factors into, then they must each start with an $x$ because the coefficient of the first term is 1.

$$(x)(x\ ) = 0$$

**3.** **Determine all the ways you can multiply to get c.**

$c = -6$, so, looking at just the positive factors, you have $1 \cdot 6$ or $2 \cdot 3$.

**4.** **Factor.**

To decide which combination should be used, look at the sign of the last term in the trinomial, the 6, which is negative. This tells you that you have to use the *difference* of two numbers in the list (think of the numbers without their signs) to get the middle term in the trinomial, the −5. In this case, one of the 1 and 6 combinations work, because their difference is 5. If you use the +1 and −6, then you get the −5 immediately from the cross-product in the FOIL process. So $(x - 6)(x + 1) = 0$.

**5.** **Use the MPZ.**

Using the MPZ, $x - 6 = 0$ or $x + 1 = 0$. This tells you that $x = 6$ or $x = -1$.

**6.** **Check.**

If $x = 6$, then $(6)^2 - 5(6) - 6 = 36 - 30 - 6 = 0$.

If $x = -1$, then $(-1)^2 - 5(-1) - 6 = 1 + 5 - 6 = 0$.

They both work!

Solve for $x$ in $6x^2 + x = 12$.

EXAMPLE

**1.** **Put the equation in the standard form.**

The first thing to do is to add −12 to each side to get the equation into the standard form for factoring and solving:

$$6x^2 + x - 12 = 0$$

This one will be a bit more complicated to factor because the 6 in the front has a couple of choices of factors, and the 12 at the end also has several choices. The trick is to pick the correct combination of choices.

**2.** **Find all the combinations that can be multiplied to get a.**

You can get 6 with $1 \cdot 6$ or $2 \cdot 3$.

**3.** **Find all the combinations that can be multiplied to get c.**

You can get 12 with $1 \cdot 12$, $2 \cdot 6$, or $3 \cdot 4$.

**4.** **Factor.**

You have to choose the factors to use so that the difference of their cross-products (outer and inner) is 1, the coefficient of the middle term. How do you know this? Because the 12 is negative, in this standard form, and the value multiplying the middle term is assumed to be 1 when there's nothing showing.

Looking this over, you can see that using the 2 and 3 from the 6 and using the 3 and 4 from the 12 will work: $2 \cdot 4 = 8$ and $3 \cdot 3 = 9$. The difference between the 8 and the 9 is, of course, 1. You can worry about the sign later.

Fill in the binomials and line up the factors so that the 2 multiplies the 4 and the 3 multiplies the 3, and you get a 6 in the front and 12 at the end. Whew!

$$(2x \quad 3)(3x \quad 4) = 0$$

The quadratic has a + on the term in the middle, so I need the bigger product of the outer and inner to be positive. I get this by making the $9x$ positive, which happens when the 3 is positive and the 4 is negative.

$$(2x + 3)(3x - 4) = 0$$

5. **Use the MPZ to solve the equation.**

The trinomial has been factored. The MPZ tells you that either $2x + 3 = 0$ or $3x - 4 = 0$. If $2x + 3 = 0$ then $2x = -3$ or $x = -\frac{3}{2}$. If $3x - 4 = 0$ then $3x = 4$ or $x = \frac{4}{3}$.

Solve for $z$ in $12z^2 - 4z - 8 = 0$.

1. **Check to see whether this quadratic is in standard form.**

You can start out by looking for combinations of factors for the 12 and the 8, but you may notice that all three terms are divisible by 4. To make things easier, take out that GCF first, and then work with the smaller numbers in the parentheses.

$$12z^2 - 4z - 8 = 4(3z^2 - z - 2) = 0$$

2. **Find the numbers that multiply to get 3.**

$$3 = 1 \cdot 3$$

3. **Find the numbers that multiply to get 2.**

$$2 = 1 \cdot 2$$

4. **Factor.**

   This is really wonderful, especially because the 3 and 2 are both prime and can be factored only one way. Your only chore is to line up the factors so there will be a difference of 1 between the cross-products.

   $$4(3z^2 - z - 2) = 4(3z \quad 2)(z \quad 1) = 0$$

   Because the middle term is negative, you need to make the larger product negative, so put the negative sign on the 1.

   $$4(3z + 2)(z - 1) = 0$$

5. **Use the MPZ to solve for the value of $z$.**

   This time, when you use the MPZ, there are three factors to consider. Either $4 = 0$, $3z + 2 = 0$, or $z - 1 = 0$. The first equation is impossible; 4 doesn't ever equal 0. But the other two equations give you answers. If $3z + 2 = 0$, then $z = -\frac{2}{3}$. If $z - 1 = 0$, then $z = 1$.

# Applying Quadratic Solutions

Quadratic equations are found in many mathematics, science, and business applications; that's why they're studied so much. The graphs of quadratic equations are always U-shaped, with an extreme point that's highest, lowest, farthest left, or farthest right. That extreme point is often the answer to a question about the situation being modeled by the quadratic. In other applications, you want the point(s) at which the U-shaped curve crosses an axis; those points are found by finding solutions to setting the quadratic equal to 0. In this section, I show you an example of how a quadratic equation is used in an application.

In physics, an equation that tells you how high an object is after a certain amount of time can be written

$$h = -16t^2 + v_0 t + h_0$$

In this equation, the $16t^2$ part accounts for the pull of gravity on the object. The number representing $v_0$ is the initial velocity — what the speed is at the very beginning. The $h_0$ is the starting height — the height in feet of the building, cliff, or stool from which the object is thrown, shot, or dropped. The variable $t$ represents time — how many seconds have passed.

**EXAMPLE**

A stone was thrown upward from the top of a 40-foot building with a beginning speed of 128 feet per second. When was the stone 296 feet up in the air?

Replacing the height, $h$, with the 296, the $v_0$ with 128, and the $h_0$ with 40, the equation now reads: $296 = -16t^2 + 128t + 40$. You can solve it using the following steps:

**1.** **Put the equation in standard form.**

Add $-296$ to each side.

$$0 = -16t^2 + 128t - 256$$

**2.** **Factor out the GCF.**

In this case, the GCF is $-16$.

$$0 = -16(t^2 - 8t + 16)$$

**3.** **Factor the quadratic trinomial inside the parentheses.**

$$0 = -16(t - 4)^2$$

**4.** **Use the MPZ to solve for the variable.**

$$t - 4 = 0, t = 4$$

After 4 seconds, the stone will be 296 feet up in the air.

## Calling on the Quadratic Formula

The quadratic formula is special to quadratic equations. A quadratic equation, $ax^2 + bx + c = 0$, can have as many as two solutions, but there may be only one solution or even no solutions at all.

**REMEMBER**

$a$, $b$, and $c$ are any real numbers. The $a$ can't equal 0, but the $b$ or $c$ can equal 0.

The quadratic formula allows you to find solutions when the equations aren't very nice. Numbers aren't *nice* when they're funky fractions, indecent decimals with no end, or raucous radicals.

**ALGEBRA RULES**

The quadratic formula says that if an equation is in the form $ax^2 + bx + c = 0$, then its solutions, the values of $x$, can be found with the following:

$$x = \frac{-b \pm \sqrt{b^2 - 4ac}}{2a}$$

You see an operation symbol, $\pm$, in the formula. The symbol is shorthand for saying that the equation can be broken into two separate equations, one using the plus sign and the other using the minus sign.

You can apply this formula to *any* quadratic equation to find the solutions — whether it factors or not. Let me show you some examples of how the formula works.

**EXAMPLE**

Use the quadratic formula to solve $2x^2 + 7x - 4 = 0$.

Refer to the standard form of a quadratic equation where the coefficient of $x^2$ is $a$, the coefficient of $x$ is $b$, and the constant is $c$. In this case, $a = 2$, $b = 7$, and $c = -4$. Inserting those numbers into the formula, you get

$$x = \frac{-7 \pm \sqrt{7^2 - 4(2)(-4)}}{2(2)}$$

Now, simplifying, and paying close attention to the order of operations, you get

$$x = \frac{-7 \pm \sqrt{49 - (-32)}}{4} = \frac{-7 \pm \sqrt{81}}{4} = \frac{-7 \pm 9}{4}$$

The two solutions are found by applying the + in front of the 9 and then the − in front of the 9.

$$x = \frac{-7 + 9}{4} = \frac{2}{4} = \frac{1}{2}$$
$$x = \frac{-7 - 9}{4} = \frac{-16}{4} = -4$$

Whenever the answers you get from using the quadratic formula come out as integers or fractions, it means that the trinomial could've been factored. It doesn't mean, though, that you shouldn't use the quadratic formula on factorable problems.

Sometimes using the quadratic formula is easier if the equation has really large or nasty numbers. In general, though, when you can, factoring using unFOIL and then the MPZ is quicker.

Just to illustrate this point, look at the previous example when it's solved using factoring and the MPZ:

$$2x^2 + 7x - 4 = (2x - 1)(x + 4) = 0$$

Then, using the MPZ, you get $2x - 1 = 0$ or $x + 4 = 0$, so $x = \frac{1}{2}$ or $x = -4$.

So, what do the results look like when the equation can't be factored? The next example shows you.

Here are two things to watch out for when using the quadratic formula:

>> **Don't forget that $-b$ means to use the *opposite* of $b$.** If the coefficient $b$ in the standard form of the equation is a positive number, change it to a negative number before inserting into the formula. If $b$ is negative, then change it to positive in the formula.

>> **Be careful when simplifying under the radical.** The order of operations dictates that you square the value of $b$ first, and then multiply the last three factors together before subtracting them from the square of $b$. Some sign errors can occur if you're not careful.

Solve for $x$ using the quadratic formula in $2x^2 + 8x + 7 = 0$.

In this problem, you let $a = 2$, $b = 8$ and $c = 7$ when using the formula:

$$x = \frac{-8 \pm \sqrt{8^2 - 4(2)(7)}}{2(2)} = \frac{-8 \pm \sqrt{64 - 56}}{4} = \frac{-8 \pm \sqrt{8}}{4}$$

The radical can be simplified because $\sqrt{8} = \sqrt{4} \cdot \sqrt{2} = 2\sqrt{2}$, so

$$x = \frac{-8 \pm 2\sqrt{2}}{4} = \frac{-^4 8 \pm 2 \sqrt{2}}{^2 4} = \frac{-4 \pm \sqrt{2}}{2}$$

Be careful when simplifying this expression: $\frac{\left(-4 + \sqrt{2}\right)}{2} \neq -2 + \sqrt{2}$. Both terms in the numerator of the fraction have to be divided by the 2.

Here are the decimal equivalents of the answers:

$$\frac{-4+\sqrt{2}}{2} \approx \frac{-4+1.414}{2} = \frac{-2.586}{2} = -1.293$$

$$\frac{-4-\sqrt{2}}{2} \approx \frac{-4-1.414}{2} = \frac{-5.414}{2} = -2.707$$

When you check these answers, what do the estimates do? If $x = -1.293$, then $2(-1.293)^2 + 8(-1.293) + 7 = 3.343698 - 10.344 + 7 = -0.000302$.

That isn't o! What happened? Is the answer wrong? No, it's okay. The rounding caused the error — it didn't come out exactly right. This happens when you use a rounded value for the answer, rather than the exact radical form. An estimate was used for the answer because the square root of a number that is not a perfect square is an irrational number, and the decimal never ends. Rounding the decimal value to three decimal places seemed like enough decimal places.

REMEMBER

You shouldn't expect the check to come out to be *exactly* o. In general, if you round the number you get from your check to the same number of places that you rounded your estimate of the radical, then you should get the o you're aiming for.

# Ignoring Reality with Imaginary Numbers

An imaginary number is something that doesn't exist — well, at least until some enterprising mathematicians had their way. Not being happy with having to halt progress in solving some equations because of negative numbers under the radical, mathematicians came up with the imaginary number $i$.

ALGEBRA
RULES

The square root of $-1$ is designated as $i$. $\sqrt{-1} = i$ and $i^2 = -1$.

Since the declaration of the value of $i$, all sorts of neat mathematics and applications have cropped up. Sorry, I can't cover all that good stuff in this book, but I at least give you a little preview of what *complex numbers* are all about.

You're apt to run into these imaginary numbers when using the quadratic formula. In the following example, the quadratic equation doesn't factor and doesn't have any *real* solutions — the only possible answers are *imaginary*.

Use the quadratic formula to solve $5x^2 - 6x + 5 = 0$.

In this quadratic, $a = 5$, $b = -6$, and $c = 5$. Putting the numbers into the formula:

$$x = \frac{-(-6) \pm \sqrt{(-6)^2 - 4(5)(5)}}{2(5)}$$

$$= \frac{6 \pm \sqrt{36 - 100}}{10}$$

$$= \frac{6 \pm \sqrt{-64}}{10}$$

You see a $-64$ under the radical. Only positive numbers and 0 have square roots. So you use the definition of the imaginary number where $i = \sqrt{-1}$ and apply it after simplifying the radical:

$$\frac{6 \pm \sqrt{-64}}{10} = \frac{6 \pm \sqrt{-1}\sqrt{64}}{10} = \frac{6 \pm i \cdot 8}{10} = \frac{3 \pm 4i}{5}$$

Applying this new *imaginary* number allowed mathematicians to finish their problems. You have two answers — although both are imaginary.

# Chapter 8

# Expanding the Equation Horizon

I n Chapters 6 and 7, you find out how to solve equations with powers of 1 and 2, and this seems to be enough to get through most of the applications. But every once in a while, you'll be thrown a curve with an equation of a degree higher than 2 or an equation with a radical in it or a fractional degree in it. No need to panic. You can deal with these rogue equations in many ways, and in this chapter, I tell you what the most efficient ways are. One common thread you'll see in solving these equations is a goal to set the expression equal to 0 so you can use the multiplication property of zero (see Chapter 7) to find the solution.

## Queuing Up to Cubic Equations

Cubic equations contain a variable term with a power of 3 but no power higher than 3. In these equations, you can expect to find up to three different solutions, but there may not be as many as three. Can you assume that fourth-degree equations could have as many as four solutions and fifth-degree equations . . . ? Yes, indeed you can — this is the general rule. The degree can tell you what the *maximum* number of solutions is. Although the number of solutions *may* be less than the number of the degree, there won't be any more solutions than that number.

# Solving perfectly cubed equations

If a cubic equation has just two terms and they're both perfect cubes, then your task is easy. The sum or difference of perfect cubes can be factored into two factors with only one solution. The first factor, or the *binomial*, gives you a solution. The second factor, the *trinomial*, does not give you a solution.

If $x^3 - a^3 = 0$, then $x^3 - a^3 = (x-a)(x^2 + ax + a^2) = 0$ and $x = a$ is the only solution. Likewise, if $x^3 + a^3 = 0$, then $(x+a)(x^2 - ax + a^2) = 0$ and $x = -a$ is the only solution. The reason you have only one solution for each of these cubics is because $x^2 + ax + a^2 = 0$ and $x^2 - ax + a^2 = 0$ have no real solutions. The trinomials can't be factored, because the quadratic formula gives you imaginary solutions.

**EXAMPLE**

Solve for $y$ in $27y^3 + 64 = 0$ using factoring.

The factorization here is $27y^3 + 64 = (3y + 4)(9y^2 - 12y + 16)$. The first factor offers a solution, so set $3y + 4$ equal to 0 to get $3y = -4$ or $y = -\frac{4}{3}$.

**EXAMPLE**

Solve for $a$ in $8a^3 - (a-2)^3 = 0$ using factoring.

The factorization here works the same as factorizations of the difference between perfect cubes. It's just more complicated because the second term is a binomial:

$$8a^3 - (a-2)^3 = [2a - (a-2)][4a^2 + 2a(a-2) + (a-2)^2] = 0$$

Simplify inside the first bracket by distributing the negative and you get

$$[2a - (a-2)] = [2a - a + 2] = [a+2]$$

Setting the first factor equal to 0, you get

$$a + 2 = 0$$
$$a = -2$$

As usual, the second factor doesn't give you a real solution, even if you distribute, square the binomial, and combine all the like terms.

# Going for the greatest common factor

Another type of cubic equation that's easy to solve is one in which you can factor out a variable greatest common factor (GCF), leaving a second factor that is linear or quadratic (first or second degree). You apply the multiplication property of zero (MPZ) and work to find the solutions — usually three of them.

## Factoring out a first-degree variable greatest common factor

When the terms of a three-term cubic equation all have the same first-degree variable as a factor, then factor that out. The resulting equation will have the variable as one factor and a quadratic expression as the second factor. The first-degree variable will always give you a solution of 0 when you apply the MPZ. If the quadratic has solutions, you can find them using the methods in Chapter 7.

Solve for $x$ in $x^3 - 4x^2 - 5x = 0$.

EXAMPLE

1. **Determine that each term has a factor of $x$ and factor that out.**

   The GCF is $x$. Factor to get $x(x^2 - 4x - 5) = 0$.

   You're all ready to apply the MPZ when you notice that the second factor, the quadratic, can be factored. Do that first and then use the MPZ on the whole thing.

2. **Factor the quadratic expression, if possible.**

   $x(x^2 - 4x - 5) = x(x - 5)(x + 1) = 0$

3. **Apply the MPZ and solve.**

   Setting the individual factors equal to 0, you get $x = 0$, $x - 5 = 0$, or $x + 1 = 0$. This means that $x = 0$ or $x = 5$ or $x = -1$.

4. **Check the solutions in the original equation.**

   If $x = 0$, then $0^3 - 4(0)^2 - 4(0)^2 - 5(0) = 0 - 0 - 0 = 0$.

   If $x = 5$, then $5^3 - 4(5)^2 - 5(5) = 125 - 4(25) - 25 = 125 - 100 - 25 = 0$.

   If $x = -1$, then $(-1)^3 - 4(-1)^2 - 5(-1) = -1 - 4(1) + 5 = -1 - 4 + 5 = 0$.

   All three work!

## Factoring out a second-degree greatest common factor

Just as with first-degree variable GCFs, you can also factor out second-degree variables (or third-degree, fourth-degree, and so on). Factoring leaves you with another expression that may have additional solutions.

Solve for $w$ in $w^3 - 3w^2 = 0$.

**1.** **Determine that each term has a factor of $w^2$ and factor that out.**

Factoring out $w^2$, you get $w^3 - 3w^2 = w^2(w-3) = 0$.

**2.** **Use the MPZ.**

$w^2 = 0$ or $w - 3 = 0$.

**3.** **Solve the resulting equations.**

Solving the first equation involves taking the square root of each side of the equation. This process usually results in two different answers — the positive answer and the negative answer. However, this isn't the case with $w^2 = 0$ because 0 is neither positive nor negative. So there's only one solution from this factor: $w = 0$. (Actually, 0 is a *double* root, because it appears twice.) And the other factor gives you a solution of $w = 3$. So, even though this is a cubic equation, there are only two unique solutions to it.

## Grouping cubes

Grouping is a form of factoring that you can use when you have four or more terms that don't have a single GCF. These four or more terms may be grouped, however, when pairs of the terms have factors in common. The method of grouping is covered in Chapter 5. I give you one example here, but turn to Chapter 5 for a more complete explanation.

Solve for $x$ in $x^3 + x^2 - 4x - 4 = 0$.

**1.** **Use grouping to factor, taking $x^2$ out of the first two terms and $-4$ out of the last two terms. Then factor $(x + 1)$ out of the newly created terms.**

$$x^3 + x^2 - 4x - 4 = x^2(x+1) - 4(x+1) = (x+1)(x^2-4) = 0$$

2. **The second factor is the difference between two perfect squares and can also be factored.**

$$(x+1)(x^2-4)=(x+1)(x-2)(x+2)=0$$

3. **Solve using the MPZ.**

$x+1=0$, $x-2=0$, or $x+2=0$, which means that $x=-1$, $x=2$, or $x=-2$.

There are three different answers in this case, but you sometimes get just one or two answers.

## Solving cubics with integers

If you can't solve a third-degree equation by finding the sum or difference of the cubes, factoring, or grouping, you can try one more method that finds all the solutions if they happen to be integers. Cubic equations could have one, two, or three different integers that are solutions. Having all three integral solutions generally only happens if the coefficient (multiplier) on the third-degree term is a 1. If the coefficient on the term with the variable raised to the third power isn't a 1, then at least one of the solutions may be a fraction (not always, but more frequently than not). Synthetic division (see the "Using Synthetic Division" section, later in this chapter) can be used to look for solutions.

Find the solutions for $x^3-7x^2+7x+15=0$ using the method of integer factors.

EXAMPLE To find the solutions when there are all integer solutions, follow these steps:

1. **Write the cubic equation in decreasing powers of the variable. Look for the constant term and list all the numbers that divide that number evenly (its factors). Remember to include both positive and negative numbers.**

   In the cubic equation $x^3-7x^2+7x+15=0$, the cubic is in decreasing powers, and the constant is 15. The list of numbers that divides 15 evenly is: ±1, ±3, ±5, and ±15. This is a long list, but you know that somehow or another the factors of the cubic have to multiply to get 15.

**2.** **Find a number from the list that makes the equation equal 0.**

Choose a 3 for your first guess. Trying $x = 3$, $(3)^3 - 7(3)^2 + 7(3) + 15 = 27 - 63 + 21 + 15 = 63 - 63 = 0$. It works!

**3.** **Divide the constant by that number.**

The answer to that division is your new constant. In the example, divide the original 15 by 3 and get 5. That's your new constant.

**4.** **Make a list of numbers that divide the new constant evenly.**

Make a new list for the new constant of 5. The numbers that divide 5 evenly are: ±1 and ±5. Four numbers are much nicer than eight.

**5.** **Find a number from the new list that checks (makes the equation equal 0).**

Trying $x = 1$, you get $(1)^3 - 7(1)^2 + 7(1) + 15 = 1 - 7 + 7 + 15 = 23 - 7 = 16$. That doesn't work, so try another number from the list.

Trying $x = 5$, $(5)^3 - 7(5)^2 + 7(5) + 15 = 125 - 175 + 35 + 15 = 175 - 175 = 0$. So, it works.

**6.** **Divide the new constant by the newest answer.**

That answer gives you the choices for the last solution.

Dividing the new constant of 5 by 5, you get 1. The only things that divide that evenly are 1 or −1. Because you already tried the 1, and it didn't work, it must mean that the −1 is the last solution.

When $x = -1$, you get $(-1)^3 - 7(-1)^2 + 7(-1) + 15 = -1 - 7 - 7 + 15 = 0$.

That does work, of course, so your solutions for $x^3 - 7x^2 + 7x + 15 = 0$ are: $x = 3$, $x = 5$, and $x = -1$. This also means that the factored version of the cubic is $(x - 3)(x - 5)(x + 1) = 0$.

Whew! That's quite a process. But it makes a lot of sense.

# Using Synthetic Division

Cubic equations that have nice integer solutions make life easier. But how realistic is that? Many answers to cubic equations that are considered to be rather nice are actually fractions. And what if you want to broaden your horizons beyond third-degree polynomials and try fourth- or fifth-degree equations or higher? Trying out guesses of answers until you find one that works can get pretty old pretty fast.

A method known as *synthetic division* can help out with all these concerns and lessen the drudgery. Synthetic division is a shortcut division process. It takes the coefficients on all the terms in an equation and provides a method for finding the answer to a division problem by only multiplying and adding. It's really pretty neat. I like to use synthetic division to help find both integer solutions and fractional solutions for polynomial equations when it's convenient and shows some promise. Refer to Chapter 5 for more on synthetic division.

Earlier in this chapter, in the "Solving cubics with integers" section, I show you how to choose possible solutions for cubic equations whose lead coefficient is a 1. This section expands your capabilities of finding rational solutions. You see how to solve equations with a degree higher than 3, and you see how to include equations whose lead coefficient is something other than 1.

Here's the general process to use:

1. Put the terms of the equation in decreasing powers of the variable.

2. List all the possible factors of the constant term.

3. List all the possible factors of the coefficient of the highest power of the variable (the *lead coefficient*).

4. Divide all the factors in Step 2 by the factors In Step 3.

   This is your list of possible *rational* solutions of the equation.

5. Use synthetic division to check the possibilities.

Find the solutions of the equation: $2x^4 + 13x^3 + 4x^2 = 61x + 30$.

**1.** **Put the terms of the equation in decreasing powers of the variable.**

$$2x^4 + 13x^3 + 4x^2 - 61x - 30 = 0$$

**2.** **List all the possible factors of the constant term.**

The constant term $-30$ has the following factors: $\pm 1, \pm 2, \pm 3,$ $\pm 5, \pm 6, \pm 10, \pm 15,$ and $\pm 30$.

**3.** **List all the possible factors of the coefficient of the highest power of the variable (the *lead coefficient*).**

The lead coefficient 2 has factors $\pm 1$ and $\pm 2$.

**4.** **Divide all the factors in Step 2 by the factors in Step 3. This is your list of possible *rational* solutions of the equation.**

Dividing the factors of $-30$ by $+1$ or $-1$ doesn't change the list of factors. Dividing by $+2$ or $-2$ adds fractions when the number being divided is odd — the even numbers just provide values already on the list. So the complete list of possible solutions is: $\pm 1, \pm 2, \pm 3, \pm 5, \pm 6, \pm 10, \pm 15, \pm 30, \pm\frac{1}{2}, \pm\frac{3}{2},$ $\pm\frac{5}{2},$ and $\pm\frac{15}{2}$.

**5.** **Use synthetic division to check the possibilities.**

I first try the number 2 as a possible solution. The final number in the synthetic division is the value of the polynomial that you get by substituting in the 2, so you want the number to be 0.

$$\underline{2|}\quad 2\quad 13\quad 4\quad -61\quad -30$$
$$\phantom{\underline{2|}\quad 2}\quad\;\; 4\quad 34\quad 76\quad\;\; 30$$
$$\overline{\phantom{\underline{2|}\;}\quad 2\quad 17\quad 38\quad\;\; 15\quad\;\;\; 0}$$

The 2 is a solution, because the final number (what you get in evaluating the expression for 2) is equal to 0.

Now look at the third row and use the lead coefficient of 2 and final entry of 15 (ignore the 0). You can now limit your choices to only factors of $+15$ divided by factors of 2. The new, revised list is: $\pm 1, \pm 3, \pm 5, \pm 15, \pm\frac{1}{2}, \pm\frac{3}{2}, \pm\frac{5}{2},$ and $\pm\frac{15}{2}$.

I would probably try only integers before trying any fractions, but I want you to see what using a fraction in synthetic division looks like. I choose to try $-\frac{1}{2}$. Use only the numbers in the last row of the previous division.

$$
\begin{array}{r|rrrr}
-\frac{1}{2} & 2 & 17 & 38 & 15 \\
& & -1 & -8 & -15 \\
\hline
& 2 & 16 & 30 & 0
\end{array}
$$

Such a wise choice! The number worked and is a solution. You could go on with more synthetic division, but, at this point, I usually stop. The first three numbers in the bottom row represent a quadratic trinomial. Write out the trinomial, factor it, use the MPZ, and find the last two solutions.

The quadratic equation represented by that last row is: $2x^2 + 16x + 30 = 0$.

First factor 2 out of each term. Then factor the trinomial: $2(x^2 + 8x + 15) = 2(x + 3)(x + 5) = 0$.

The solutions from the factored trinomial are $x = -3$ and $x = -5$. Add these two solutions to $x = 2$ and $x = -\frac{1}{2}$, and you have the four solutions of the polynomial.

What? You're miffed! You wanted me to finish the problem using synthetic division — not bail out and factor? Okay. I'll pick up where I left off with the synthetic division and show you how it finishes:

$$
\begin{array}{r|rrr}
-3 & 2 & 16 & 30 \\
& & -6 & -30 \\
\hline
& 2 & 10 & 0
\end{array}
$$

And, finally:

$$
\begin{array}{r|rr}
-5 & 2 & 10 \\
& & -10 \\
\hline
& 2 & 0
\end{array}
$$

# Working Quadratic-Like Equations

Some equations with higher powers or fractional powers are *quadratic-like*, meaning that they have three terms and

» The variable in the first term has an even power (4, 6, 8, . . .) or $\left(\frac{1}{2}, \frac{1}{4}, \frac{1}{6}, \ldots\right)$.

» The variable in the second term has a power that is half that of the first.

» The third term is a constant number.

In general, the format for a quadratic-like equation is: $ax^{2n} + bx^{n} + c = 0$. Just as in the general quadratic equation, the $x$ is the variable and the $a$, $b$, and $c$ are constant numbers. The $a$ can't be 0, but the other two letters have no restrictions. (I show you how to factor quadratic-like expressions in Chapter 5.)

To solve a quadratic-like equation, first pretend that it's quadratic, and use the same methods as you do for those; then do a step or two more. The extra steps usually involve taking an extra root or raising to an extra power.

Now I'll show you the steps used to solve quadratic-like equations by working through a couple of examples.

Solve for $x$ in $x^4 - 5x^2 + 4 = 0$.

**EXAMPLE**

**1.** **Rewrite the equation, replacing the actual powers with the numbers 2 and 1.**

Rewrite this as a quadratic equation using the same *coefficients* (number multipliers) and constant.

**TIP**

Change the letter used for the variable, so you won't confuse this new equation with the original. Substitute $q$ for $x^2$ and $q^2$ for $x^4$:

$$q^2 - 5q + 4 = 0$$

**2.** **Factor the quadratic equation.**

$q^2 - 5q + 4 = 0$ factors nicely into $(q - 4)(q - 1) = 0$.

3. **Reverse the substitution and use the factorization pattern to factor the original equation.**

   Use that same pattern to write the factorization of the original problem. When you replace the variable $q$ in the factored form, use $x^2$:

   $$x^4 - 5x^2 + 4 = (x^2 - 4)(x^2 - 1) = 0$$

4. **Solve the equation using the MPZ.**

   Either $x^2 - 4 = 0$ or $x^2 - 1 = 0$. If $x^2 - 4 = 0$, then $x^2 = 4$ and $x = \pm 2$. If $x^2 - 1 = 0$, then $x^2 - 1$ and $x^2 = \pm 1$.

This fourth-degree equation did live up to its reputation and have four different solutions.

This next example presents an interesting problem because the exponents are fractions. But the trinomial fits into the category of *quadratic-like*, so I'll show you how you can take advantage of this format to solve the equation. And, no, the rule of the number of solutions doesn't work the same way here. There aren't any possible situations where there's half a solution.

Solve $w^{\frac{1}{2}} - 7w^{\frac{1}{4}} + 12 = 0$.

EXAMPLE

1. **Rewrite the equation with powers of 2 and 1. Substitute $q$ for $w^{\frac{1}{4}}$ and $q^2$ for $w^{\frac{1}{2}}$.**

   *Remember:* Squaring $w^{\frac{1}{4}}$ gives you $\left(w^{\frac{1}{4}}\right)^2 = w^{\frac{2}{4}} = w^{\frac{1}{2}}$.

   Rewrite the equation as $q^2 - 7q + 12 = 0$.

2. **Factor.**

   This factors nicely into $(q - 3)(q - 4) = 0$.

3. **Replace the variables from the original equation, using the pattern.**

   Replace with the original variables to get $\left(w^{\frac{1}{4}} - 3\right)\left(w^{\frac{1}{4}} - 4\right) = 0$.

4. **Solve the equation for the original variable, $w$.**

   $$\left(w^{\frac{1}{4}} - 3\right)\left(w^{\frac{1}{4}} - 4\right) = 0$$

   Now, when you use the MPZ, you get that either $w^{\frac{1}{4}} - 3 = 0$ or $w^{\frac{1}{4}} - 4 = 0$. How do you solve these things? Look at $w^{\frac{1}{4}} - 3 = 0$. Adding 3 to each side, you get $w^{\frac{1}{4}} = 3$. You can solve for $w$ if you raise each side to the fourth power: $\left(w^{\frac{1}{4}}\right)^4 = (3)^4$. This says that $w = 81$. Doing the same with the other factor, if $w^{\frac{1}{4}} - 4 = 0$, then $w^{\frac{1}{4}} = 4$ and $\left(w^{\frac{1}{4}}\right)^4 = (4)^4$. This says that $w = 256$.

5. **Check the answers.**

If $w = 81, (81)^{\frac{1}{2}} - 7(81)^{\frac{1}{4}} + 12 = 9 - 7(3) + 12 = 21 - 21 = 0$.

If $w = 256, (256)^{\frac{1}{2}} - 7(256)^{\frac{1}{4}} + 12 = 16 - 7(4) + 12 = 28 - 28 = 0$.

They both work.

Negative exponents are another interesting twist to these equations, as you see in the next example.

**EXAMPLE**

Solve for the value of $x$ in $2x^{-6} - x^{-3} - 3 = 0$.

1. **Rewrite the equation using powers of 2 and 1. Substitute $q$ for $x^{-3}$ and $q^2$ for $x^{-6}$.**

   Rewrite the equation as $2q^2 - q - 3 = 0$.

2. **Factor.**

   This factors into $(2q - 3)(q + 1) = 0$.

3. **Go back to the original variables and powers.**

   Use this pattern. Factor the original equation to get:

   $$(2x^{-3} - 3)(x^{-3} + 1) = 0$$

4. **Solve.**

   Use the MPZ. The two equations to solve are $2x^{-3} - 3 = 0$ and $x^{-3} + 1 = 0$. These become $2x^{-3} = 3$ and $x^{-3} = -1$. Rewrite these using the definition of negative exponents:

   $$x^{-n} = \frac{1}{x^n}$$

   So the two equations can be written $\frac{2}{x^3} = 3$ and $\frac{1}{x^3} = -1$.

   Cross-multiply in each case to get $3x^3 = 2$ and $x^3 = -1$. Divide the first equation through by 3 to get the $x^3$ alone, and then take the cube root of each side to solve for $x$:

   $$x = \sqrt[3]{\frac{2}{3}} \text{ or } x = \sqrt[3]{-1} = -1$$

# Rooting Out Radicals

Some equations have radicals in them. You change those equations to linear or quadratic equations for greater convenience when solving. A basic process that leads to a solution of equations involving a radical involves getting rid of that radical. Removing the radical changes the problem into something more manageable, but the change also introduces the possibility of a nonsense answer or an error. Checking your answer is even more important in the case of solving radical equations.

The main method to use when dealing with equations that contain radicals is to change the equations to those that do not have radicals in them. You accomplish this by raising the radical to a power that changes the fractional exponent (representing the radical) to a 1.

Raising to powers clears out the radicals, but problems can occur when the variables are raised to even powers. Variables can stand for negative numbers or values that allow negatives under the radical, which isn't always apparent until you get into the problem and check an answer. Instead of going on with all this doom and gloom and the problems that occur when powering up both sides of an equation, let me show you some examples of how the process works, what the pitfalls are, and how to deal with any extraneous solutions.

Solve for $x$ in $2\sqrt{x+15} - 3 = 9$.

**EXAMPLE**

**1.** **Get the radical term by itself on one side of the equation.**

The first step is to add 3 to each side: $2\sqrt{x+15} = 12$.

**2.** **Square both sides of the equation. (You could divide both sides by 2, but I want to show you an important rule when squaring both sides.)**

**ALGEBRA RULES**

One of the rules involving exponents is the square of the product of two factors is equal to the product of each of those same factors squared: $(a \cdot b)^2 = a^2 \cdot b^2$.

Squaring the left side,

$$\left(2\sqrt{x+15}\right)^2 = 2^2 \cdot \left(\sqrt{x+15}\right)^2 = 4(x+15).$$

Squaring the right side, $12^2 = 144$.

So you get the new equation: $4(x+15) = 144$.

**3.** **Solve for x in the new, linear equation.**

Distribute the 4, first: $4x + 60 = 144$.

Subtract 60 from each side, and you get $4x = 84$ or $x = 21$.

**4.** **Check your work.**

$$2\sqrt{x+15} - 3 = 9$$
$$2\sqrt{21+15} - 3 = 2\sqrt{36} - 3$$
$$= 2 \cdot 6 - 3$$
$$= 12 - 3$$
$$= 9$$

Next I show you an example where you find two different solutions, but only one of them works.

Solve for $z$ in $7 + \sqrt{z-1} = z$.

EXAMPLE **1.** **Get the radical by itself on the left.**

Subtracting 7 from each side, you end up with the radical on the left and a binomial on the right.

$$\sqrt{z-1} = z - 7$$

**2.** **Square both sides of the equation.**

The only thing to watch out for here is squaring the binomial correctly.

$$\left(\sqrt{z-1}\right)^2 = (z-7)^2$$
$$z - 1 = z^2 - 14z + 49$$

**3.** **Solve the equation.**

This time you have a quadratic equation. Move everything over to the right, so that you can set the equation equal to 0. To do this, subtract $z$ from each side and add 1 to each side.

$$0 = z^2 - 15z + 50$$

The right side factors, giving you $(z-5)(z-10) = 0$. Using the MPZ, you get either $z - 5$ or $z - 10$.

**4.** **Check your answer.**

Check these carefully because incorrect answers often show up — especially when you create a quadratic equation by the squaring-each-side process.

If $z - 5$, then $7 + \sqrt{5-1} = 7 + \sqrt{4} = 7 + 2 = 9 \neq 5$. The 5 doesn't work.

If $z - 10$, then $7 + \sqrt{10-1} = 7 + \sqrt{9} = 7 + 3 = 10$. The 10 does work.

The only solution is that $z$ equals 10. That's fine. Sometimes these problems have two answers, sometimes just one answer, or sometimes no answer at all. The method works — you just have to be careful.

# Chapter 9

# Reconciling Inequalities

E quality is an important tool in mathematics and science. This chapter introduces you to algebraic *inequality*, which isn't exactly the opposite of equality. You could say that algebraic inequality is a bit like equality but softer. You use inequality for comparisons — when you're determining if something is positive or negative, bigger than or smaller than, between numbers, or infinite. Inequality allows you to sandwich expressions between values on the low end and the high end.

Algebraic inequalities show relationships between a number and an expression or between two expressions. The inequality relation is a bit less than precise. One thing can be bigger by a lot or bigger by a little, but there's still that relationship between them — that one is bigger than the other.

Many operations involving inequalities work the same as operations on equalities and equations, but you need to pay attention to some important differences that I show you in this chapter.

# Introducing Interval Notation

Algebraic operations and manipulations are performed on inequality statements while they're in an inequality format. You see the inequality statements written using the following notation:

» $<$: Less than

» $>$: Greater than

» $\leq$: Less than or equal to

» $\geq$: Greater than or equal to

To keep the direction straight as to which way to point the arrow, just remember that the itsy-bitsy part of the arrow is next to the smaller (itsy-bitsier) of the two values.

Inequality statements have been around for a long time. The symbols are traditional and accepted by mathematicians around the world. But a new notation is gaining popularity, called *interval notation*. Interval notation uses parentheses and brackets instead of inequality symbols, and it introduces the infinity symbol.

## Comparing inequality and interval notation

Before defining how interval notation is used, let me first write the same statement in both inequality and interval notation:

| Inequality | Interval Notation |
|---|---|
| $x > 8$ | $(8, \infty)$ |
| $x < 2$ | $(-\infty, 2)$ |
| $x \geq -7$ | $[-7, \infty)$ |
| $x \leq 5$ | $(-\infty, 5]$ |
| $-4 < x \leq 10$ | $(-4, 10]$ |

So, now that you've seen interval notation in action, let me give you the rules for using it.

**ALGEBRA RULES**

Interval notation expresses inequality statements with the following rules:

>> Parentheses to show *less than* or *greater than* (but not including)

>> Brackets to show *less than or equal to* or *greater than or equal to*

>> Parentheses at both infinity or negative infinity

>> Numbers and symbols written in the same left-to-right order as on a number line

**EXAMPLE**

Here are some examples of writing inequality statements using interval notation or vice versa:

>> $-3 \leq x \leq 11$ becomes $[-3, 11]$.

>> $-4 \leq x < -3$ becomes $[-4, -3)$.

>> $x > -9$ becomes $(-9, \infty)$.

>> $5 < x$ becomes $(5, \infty)$. Notice that the variable didn't come first in the inequality statement, and saying 5 must be smaller than some numbers is the same as saying that those numbers are bigger (greater) than 5, or $x > 5$.

>> $4 < x < 15$ becomes $(4, 15)$. Here's my biggest problem with interval notation: The notation $(4, 15)$ looks like a point on the coordinate plane, not an interval containing numbers between 4 and 15. You just have to be aware of the context when you come across this notation.

>> $[-8, 5]$ becomes $-8 \leq x \leq 5$.

>> $(-\infty, 0]$ becomes $x \leq 0$.

>> $(44, \infty)$ becomes $x > 44$.

## Graphing inequalities

One of the best ways of describing inequalities is with a graph. Graphs in the form of number lines are a great help when solving quadratic inequalities (see the "Taking on Quadratic and Rational Inequalities" section, later in this chapter).

A number-line graph of an inequality consists of numbers representing the starting and ending points of any interval described by the inequality and symbols above the numbers indicating whether the number is to be included in the answer. The symbols used with inequality notation are hollow circles and filled-in circles. The symbols used with interval notation are the same parentheses and brackets used in the statements.

**EXAMPLE**

Write the statement "all numbers between –3 and 4, including the 4" in inequality notation and interval notation. Then graph the inequality using both types of notation.

>> The inequality notation is $-3 < x \le 4$. The graph is shown in Figure 9-1.

>> The interval notation is $(-3, 4]$. The graph is shown in Figure 9-2.

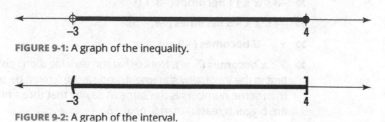

**FIGURE 9-1:** A graph of the inequality.

**FIGURE 9-2:** A graph of the interval.

# Performing Operations on Inequalities

There are many similarities between working with inequalities and working with equations. The balancing part still holds. It's when operations like multiplying each side by a number or dividing each side by a number come into play that there are some differences.

**ALGEBRA RULES**

The rules for operations on inequalities are given here. I'm showing the rules only for less than (<), but they also apply to greater than (>):

>> If $a < b$, then $a + c < b + c$ and $a - c < b - c$. The direction of the inequality stays the same.

>> If $a < b$ and $c$ is positive, then $a \cdot c < b \cdot c$ and $\dfrac{a}{c} < \dfrac{b}{c}$. The direction of the inequality stays the same.

>> If $a < b$ and $c$ is negative, then $a \cdot c > b \cdot c$ and $\dfrac{a}{c} > \dfrac{b}{c}$. When multiplying or dividing with a negative number, the direction of the inequality symbol changes.

>> If $\dfrac{a}{c} < \dfrac{b}{d}$, then $\dfrac{c}{a} > \dfrac{d}{b}$. The inequality symbol changes when you *flip* (write the reciprocals of) the fractions.

# Adding and subtracting numbers to inequalities

Adding and subtracting values within inequalities works exactly the same as with equations. You keep things balanced. Let me show you how this works.

Start with an inequality statement that you can tell is true by looking at it, such as 6 is less than 10:

$$6 < 10$$

What happens if you add the same thing to each side? You can do that to an equation and not have the truth change, but what about an inequality? Add 4 to each side:

$$6 + 4 < 10 + 4$$
$$10 < 14$$

Ten is less than 14 is still a true statement. This demonstration isn't enough to prove anything, but it does illustrate a rule that is true: When you add any number to both sides of an inequality, the inequality is still correct or true. Similarly, when you subtract any number from both sides of an inequality, the inequality is still correct or true.

# Multiplying and dividing inequalities

Now come the tricky operations. Multiplication and division add a new dimension to working with inequalities.

When multiplying or dividing both sides of an inequality by a positive number, the inequality remains correct or true. When multiplying or dividing both sides by a negative number, the inequality sign has to be reversed — point in the opposite direction — for the inequality to be correct or true. You can never multiply each side by 0 — that always makes it false (unless you have an *or equal to*). And, of course, you can never divide anything by 0.

Start with positive numbers, such as 20 and 12:

$$20 > 12$$

Multiply each side by 4:

$$20 \cdot 4 > 12 \cdot 4$$
$$80 > 48$$

It's still true. So is there a problem?

You can see the complication with my new inequality, $10 > -3$. Multiply each side by $-2$:

$$10(-2) > -(-2)$$
$$-20 > 6$$

Oops! A negative can't be greater than a positive:

$$-20 < 6$$

Making the inequality untrue is bad news. The good news is that turning the inequality symbol around is a relatively easy way to fix this.

**REMEMBER**

Whenever you multiply each side of an inequality by a negative number (or divide by a negative number), turn the inequality symbol to face the opposite direction.

**WARNING**

In the case of inequalities, you can neither divide nor multiply by 0. Of course, dividing by 0 is always forbidden, but you can usually multiply expressions by 0 (and get a product of 0). However, you can't multiply inequalities by 0.

Look at what happens when each side of an inequality is multiplied by 0:

$$3 < 7$$
$$0 \cdot 3 < 0 \cdot 7$$
$$0 < 0$$

No! It's just not true: Zero is not less than itself, nor is it greater than itself. So, to keep 0 from getting an inferiority or superiority complex, don't use it to multiply inequalities.

# Finding Solutions for Linear Inequalities

Linear inequalities, like linear equations, are those statements in which the exponent on the variable is no more than 1. Solving linear inequalities is much like solving linear equations. The main thing to remember is to reverse the inequality symbol when you multiply or divide by a negative number — and only then. You also need to keep in mind that you don't get just a single answer to linear inequalities but a whole bunch of answers — an infinite number of answers.

**EXAMPLE**

Solve for the values of $z$ in $-2(3z+4) > 10$.

In this case, the only variable term is already on the left. A usual next step would be to distribute the $-2$ over the terms on the left. But, because 2 divides 10 evenly, an alternate step lets you avoid having to do the distribution.

**1.** **Divide each side by $-2$.**

Be sure to switch the inequality symbol around.

$$\frac{-2(3z+4)}{-2} < \frac{10}{-2}$$

$$3z+4 < -5$$

**2.** **Subtract 4 from each side.**

$$3z+4-4 < -5-4$$

$$3z < -9$$

**3.** **Divide each side by 3.**

$$\frac{3z}{3} < \frac{-9}{3}$$

$$z < -3$$

# Expanding to More than Two Expressions

One big advantage that inequalities have over equations is that they can be expanded or strung out into compound statements, and you can do more than one comparison at the same time. Look at this statement:

$$2 < 4 < 7 < 11 < 12$$

You can create another true statement by pulling out any pair of numbers from the inequality, as long as you write them in the same order. They don't even have to be next to one another. For example:

$$4 < 12 \qquad 2 < 11 \qquad 2 < 12$$

One thing you can't do, though, is to mix up inequalities, going in opposite directions, in the same statement. You can't write $7 < 12 < 2$.

The operations on these compound inequality expressions use the same rules as for the linear expressions (refer to the "Performing Operations on Inequalities" section, earlier in this chapter). You just extend the process by performing the operations on each section or part.

Solve for the values of $x$ in $-3 \le 5x + 2 < 17$.

1.  **The goal is to get the variable alone in the middle. Start by subtracting 2 from each section.**

    $-3 - 2 \le 5x + 2 - 2 < 17 - 2$

    $-5 \le 5x < 15$

2.  **Now divide each section by 5.**

    The number 5 is positive, so don't turn the inequality signs around.

    $$\frac{-5}{5} \le \frac{5x}{5} < \frac{15}{5}$$

    $-1 \le x < 3$

    This says that $x$ is greater than or equal to $-1$ while, at the same time, it's less than 3. Some possible solutions are: 0, 1, 2, 2.9.

3.  **Check the solution using two of these possibilities.**

    If $x = 1$, then $-3 \le 5(1) + 2 < 17$, or $-3 \le 7 < 17$. That's true.

    If $x = 2$, then $-3 \le 5(2) + 2 < 17$, or $-3 \le 12 < 17$. This also works.

# Taking on Quadratic and Rational Inequalities

A *quadratic inequality* is an inequality that involves a variable term with a second-degree power (and no higher powers of the variable). When solving quadratic inequalities, the rules of addition, subtraction, multiplication, and division of inequalities still hold, but the final step in the solution is different. The best way to describe how to solve a quadratic inequality is to use an example and put the rules right in the example.

Solve for $x$ in $x^2 + 3x > 4$.

**EXAMPLE**

**1. Move all terms to one side.**

First, move the 4 to the left by subtracting 4 from each side.

$$x^2 + 3x > 4$$
$$x^2 + 3x - 4 > 0$$

**2. Factor.**

Factor the quadratic on the left using unFOIL.

$$(x+)(x-1) > 0$$

**3. Find all the values of $x$ that make the factored side equal to 0.**

In this case, there are two values. Using the multiplication property of zero, you get $x + 4 = 0$ or $x - 1 = 0$, which results in $x = -4$ or $x = 1$.

The *multiplication property of zero* says that if the product of two or more factors equals 0, then at least one of the factors must be zero. If $xyz = 0$, then $x = 0$ or $y = 0$ or $z = 0$.

**REMEMBER**

**4. Make a number line listing the values from Step 3, and determine the signs of the expression between the values on the chart.**

When you choose a number to the left of $-4$, both factors are negative, and the product is positive. Between $-4$ and 1, the first factor is positive and the second factor is negative, resulting in a negative product. To the right of 1, both factors are positive, giving you a positive product. Just testing one of the numbers in the interval tells you what will happen to all of them.

Figure 9-3 shows you a number line with the critical numbers in their places and the signs in the intervals between the points.

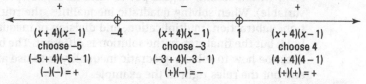

**FIGURE 9-3:** A number line helps you find the signs of the factors and their products.

5. **Determine which intervals give you solutions to the problem.**

The values for $x$ that work to make the quadratic $x^2 + 3x - 4 > 0$ are all the negative numbers smaller than –4 down lower to really small numbers and all the positive numbers bigger than 1 all the way up to really big numbers. The only numbers that don't work are those between –4 and 1. You write your answer as $x < -4$ or $x > 1$.

In interval notation, the answer is $(-\infty, -4) \cup (1, \infty)$. The $\cup$ symbol is for *union*, meaning everything in either interval (one or the other) works.

## Using a similar process with more than two factors

Even though this section involves problems that are *quadratic inequalities* (inequalities that have at least one squared variable term and a greater-than or less-than sign), some other types of inequalities belong in the same section because you handle them the same way as you do quadratics. You can really have any number of factors and any arrangement of factors and do the positive-and-negative business to get the answer, as I show you in the following example.

EXAMPLE

Solve for the values of $x$ that work in $(x-4)(x+3)(x-2)(x+7) > 0$.

This problem is already factored, so you can easily determine that the numbers that make the expression equal to 0 (the critical numbers) are $x = 4$, $x = -3$, $x = 2$, $x = -7$. Put them in order from the smallest to the largest on a number line (see Figure 9-4), and test for the signs of the products in the intervals.

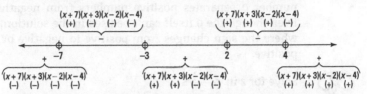

**FIGURE 9-4:** The sign changes at each critical number in this problem.

Because the original problem is looking for values that make the expression greater than 0, or positive, the solution includes numbers in the intervals that are positive. Those numbers are

>> Smaller than $-7$

>> Between $-3$ and $2$

>> Bigger than $4$

The solution is written $x < -7$ or $-3 < x < 2$ or $x > 4$. In interval notation, the solution is written $(-\infty, -7) \cup (-3, 2) \cup (4, \infty)$.

## Identifying the factors in fractional inequalities

Inequalities with fractions that have variables in the denominator are another special type of inequality that fits under the general heading of quadratic inequalities; they get to be in this chapter because of the way you solve them.

To solve these rational (fractional) inequalities, do somewhat the same thing as you do with the inequalities dealing with two or more factors:

1. **Find where the expression equals 0.**

   Actually, expand that to looking for, separately, what makes the numerator (top) equal to 0 and what makes the denominator (bottom) equal to 0. These are your *critical numbers.*

2. **Check the intervals between the critical numbers.**

3. **Write the answer.**

The one big caution with rational inequalities is not to include any number in the final answer that makes the denominator of the fraction equal 0. Zero in the denominator makes it an impossible situation, not to mention an impossible fraction.

So why look at what makes the denominator 0 at all? The number 0 separates positive numbers from negative numbers. Even though the 0 itself can't be used in the solution, it indicates where the sign changes from positive to negative or negative to positive.

EXAMPLE

Solve for $z$ in $\dfrac{z^2-1}{z^2-9} \le 0$.

Factor the numerator and denominator to get $\dfrac{(z+1)(z-1)}{(z+3)(z-3)} \le 0$.

The numbers making the numerator or denominator equal to 0 are $z = +1, -1, +3, -3$. Make a number line that contains the critical numbers and the signs of the intervals (see Figure 9-5).

| + | | − | | + | | − | | + |
|---|---|---|---|---|---|---|---|---|
| $\dfrac{(z+1)(z-1)}{(z+3)(z-3)}$ | −3 | $\dfrac{(z+1)(z-1)}{(z+3)(z-3)}$ | −1 | $\dfrac{(z+1)(z-1)}{(z+3)(z-3)}$ | 1 | $\dfrac{(z+1)(z-1)}{(z+3)(z-3)}$ | 3 | $\dfrac{(z+1)(z-1)}{(z+3)(z-3)}$ |
| $\dfrac{(-)(-)}{(-)(-)}$ | | $\dfrac{(-)(-)}{(+)(-)}$ | | $\dfrac{(+)(-)}{(+)(-)}$ | | $\dfrac{(+)(+)}{(+)(-)}$ | | $\dfrac{(+)(+)}{(+)(+)}$ |

**FIGURE 9-5:** The 1 and −1 are included in the solution.

Because you're looking for values of $z$ that make the expression negative, you want the values between −3 and −1 and those between 1 and 3. Also, you want values that make the expression equal to 0. That can only include the numbers that make the numerator equal to 0, the 1 and −1. The answer is written

$$-3 < z \le -1 \text{ or } 1 \le z < 3$$

In interval notation, the solution is written

$$(-3, -1] \cup [1, 3)$$

Notice that the $<$ symbol is used by the −3 and 3 so those two numbers don't get included in the answer.

IN THIS CHAPTER

» Changing from an absolute value
equation to separate linear equations

» Recognizing when no solution is possible

» Transforming an absolute value
inequality into one or two statements

# Chapter **10**

# Absolute-Value Equations and Inequalities

The *absolute value* function actually measures a distance. How far is the number from 0? So the direction of a value — right or left of zero — doesn't make any difference in the world of absolute value. The symbol that signifies that you're performing the absolute-value function is two vertical lines — you sandwich the number to be operated upon between the lines. Absolute value strips away negative signs. Because of this, when solving equations or inequalities involving absolute value, you have to account for the original number having been either positive or negative.

## Acting on Absolute-Value Equations

Before tackling the inequalities, take a look at absolute-value equations. An equation such as $|x| = 7$ is fairly easy to decipher. It's asking for values of $x$ that give you a 7 when you put it in the absolute-value symbol. Two answers, 7 and −7, have an absolute value of 7. Those are the only two answers. But what about something a bit more involved, such as $|3x + 2| = 4$? The equation is true if the sum of $3x$ and 2 is equal to +4. But it's also true if the sum of $3x$ and 2 is equal to −4. The two possibilities for the sum result in two possibilities for the value of $x$.

To solve an absolute-value equation of the form $|ax + b| = c$, change the absolute-value equation to two equivalent linear equations and solve them.

$|ax + b| = c$ is equivalent to $ax + b = c$ or $ax + b = -c$. Notice that the left side is the same in each equation. The $c$ is positive in the first equation and negative in the second because the expression inside the absolute-value symbol can be positive or negative — absolute value makes them both positives when it's performed.

Solve for $x$ in $|3x + 2| = 4$.

1.  **Rewrite as two linear equations.**

    $3x + 2 = 4$ or $3x + 2 = -4$

2.  **Solve for the value of the variable in each of the equations.**

    Subtract 2 from each side in each equation: $3x = 2$ or $3x = -6$.

    Divide each side in each equation by 3: $x = \frac{2}{3}$ or $x = -2$.

3.  **Check.**

    If $x = -2$, then $|3(-2) + 2| = |-6 + 2| = |-4| = 4$.

    If $x = \frac{2}{3}$, then $\left|3\left(\frac{2}{3}\right) + 2\right| = |2 + 2| = 4$.

    They both work.

In the next example, you see the equation set equal to 0. For these problems, though, you don't want a number added to or subtracted from the absolute value on the same side of the equal sign. In order to use the rule for changing to linear equations, you have to have the absolute value by itself on one side of the equation.

Solve for $x$ in $|5x - 2| + 3 = 0$.

1. **Get the absolute-value expression by itself on one side of the equation.**

   Adding –3 to each side:

   $$|5x - 2| = -3$$

2. **Rewrite as two linear equations.**

   $5x - 2 = -3$ or $5x - 2 = +3$

3. **Solve the two equations for the value of the variable.**

   Add 2 to each side of the equations:

   $5x = -1$ or $5x = 5$

   Divide each side by 5:

   $$x = -\frac{1}{5} \text{ or } x = 1$$

4. **Check.**

   If $x = -\frac{1}{5}$ then, $\left|5\left(-\frac{1}{5}\right) - 2\right| + 3 = |-1 - 2| + 3 = |-3| + 3 = 6$.

   Oops! That's supposed to be a 0. Try the other one.

   If $x = 1$, then $|5(1) - 2| + 3 = |3| + 3 = 6$.

   No, that didn't work either.

Now's the time to realize that the equation was impossible to begin with. (Of course, noticing this before you started would've saved time.) The definition of absolute value tells you that it results in everything being positive. Starting with an absolute value equal to –3 gave you an impossible situation to solve. No wonder you didn't get an answer!

# Working Absolute-Value Inequalities

Absolute-value inequalities are just what they say they are — inequalities that have absolute-value symbols somewhere in the problem.

REMEMBER

$|a|$ is equal to $a$ if $a$ is a positive number or 0. $|a|$ is equal to the opposite of $a$, or $-a$, if $a$ is a negative number. So $|3| = 3$ and $|-7| = -(-7) = 7$.

Absolute-value equations and inequalities can look like the following:

$$|x+3| = 5 \qquad |2x+3| > 7 \qquad |5x+1| \le 9$$

Solving absolute-value inequalities brings two different procedures together into one topic. The first procedure involves the methods similar to those used to deal with absolute-value equations, and the second involves the rules used to solve inequalities. You might say it's the best of both worlds. Or you might not.

ALGEBRA RULES

To solve an absolute-value inequality of the form $|ax+b| > c$, change the absolute-value inequality to two inequalities equivalent to that original problem and solve them: $|ax+b| > c$ is equivalent to $ax+b > c$ or $ax+b < -c$. Notice that the inequality symbol is reversed with the $-c$.

Solve for $x$ in $|2x-5| > 7$.

EXAMPLE

**1.** **Rewrite as two inequalities.**

$2x-5 > 7$ or $2x-5 < -7$

**2.** **Solve each inequality.**

Add 5 to each side in each inequality:

$2x > 12$ or $2x < -2$

Divide through by 2:

$x > 6$ or $x < -1$

In interval notation, that's $(-\infty, -1) \cup (6, \infty)$. (See Chapter 9 for more on interval notation.)

The answer seems to go in two different directions — and it does. You need numbers that get larger and larger to keep the result bigger than 7, and you need numbers that get smaller and smaller so that the absolute value of the small negative numbers is also bigger than 7. That's why, when doing the solving, you use both greater than the $+c$ and less than the $-c$ in the problem.

Now, consider the absolute-value inequality that is kept small. The result of performing the absolute value can't be too large — it has to be smaller than $c$.

To solve an absolute-value inequality of the form $|ax+b|<c$, change the absolute-value inequality to an equivalent compound inequality and solve it: $|ax+b|<c$ is equivalent to $-c<ax+b<c$.

Solve for $x$ in $|5x+1|\le 9$.

**1.** **Rewrite as two inequalities.**

$-9\le 5x+1\le 9$

**2.** **Solve the inequality.**

Subtract 1 from each section:

$-10\le 5x\le 8$

Now divide through by 5:

$-2\le x\le \frac{8}{5}$ or, in interval notation, $\left[-2,\frac{8}{5}\right]$

Notice that this problem had a less-than-or-equal-to symbol. The rules for *less than* or *greater than* are the same as those for the problems including the endpoints of the interval — when the numbers establishing the starting or ending points are included in the answer.

# Chapter **11**

# Making Algebra Tell a Story

S tory problems can be one of the least favorite activities for algebra students. Although algebra and its symbols, rules, and processes act as a door to higher mathematics and logical thinking, story problems give you immediate benefits and results in real-world terms.

Algebra allows you to solve problems. Not all problems — it won't help with that noisy neighbor — but problems involving how to divvy up money equitably or make things fit in a room. In this chapter, you find some practical applications for algebra. You may not be faced with the exact situations I use in this chapter, but you should find some skills that will allow you to solve the story problems or practical applications that are special to your situation.

## Making Plans to Solve Story Problems

When solving story problems, the equation you should use or how all the ingredients interact isn't always immediately apparent. It helps to have a game plan to get you started. Sometimes, just picking up a pencil and drawing a picture can be a big help. Other times, you can just write down all the numbers involved.

**TIP**

You don't have to use every suggestion in the following list with every problem, but using as many as possible can make the task more manageable:

» **Draw a picture.** Label your picture with numbers or names or other information that helps you make sense of the situation. Fill it in more or change the drawing as you set up an equation for the problem.

» **Assign a variable(s) to represent *how many* or *number of*.** You may use more than one variable at first and refine the problem to just one variable later.

» **If you use more than one variable, go back and substitute known relationships for the extra variables.** When it comes to solving the equations, you want to solve for just one variable. You can often rewrite all the variables in terms of just one of them.

» **Look at the end of the question or problem statement.** This often gives a big clue as to what's being asked for and what the variables should represent. It can also give a clue as to what formula to use, if a formula is appropriate.

» **Translate the words into an equation.** Replace:
   • and, more than, and exceeded by with the plus sign (+)
   • less than, less, and subtract from with the minus sign (−)
   • of and times as much with the multiplication sign (×)
   • twice with two times (2 ×)
   • divided by with the division sign (÷)
   • half as much with one-half times $\left(\frac{1}{2} \times\right)$
   • the verb (is or are, for example) with the equal sign (=)

» **Plug in a standard formula, if the problem lends itself to one.** When possible, use a formula as your equation or as part of your equation. Formulas are a good place to start to set up relationships. Be familiar with what the variables in the formula stand for.

» **Check to see if the answer makes any sense.** When you get an answer, decide whether it makes sense within the context of the problem. Having an answer make sense doesn't guarantee that it's a correct answer, but it's the first check to tell if it isn't correct.

# Finding Money and Interest Interesting

Figuring out how much interest you have to pay, or how much you're earning in interest, is simple with the proper formulas. And, when the formulas are incorporated into story problems, the processes just seem to fall into place.

**ALGEBRA RULES**

The amount of simple interest earned is equal to the amount of the principal, $P$ (the starting amount), times the rate of interest, $r$ (which is written as a decimal), times the amount of time, $t$ (usually in years). The formula to calculate simple interest is: $I = Prt$.

What is the amount of simple interest on \$10,000 when the interest rate is $2\frac{1}{2}$ percent and the time period is $3\frac{1}{2}$ years?

$$I = Prt$$
$$I = 10,000 \cdot 0.025 \cdot 3.5 = 875$$

The interest is \$875.

See! That was a *simple* story problem!

## Investigating investments and interest

You can invest your money in a safe CD or savings account and get one interest rate. You can also invest in riskier ventures and get a higher interest rate, but you risk losing money. Most financial advisors suggest that you diversify — put some money in each type of investment — to take advantage of each investment's good points.

Use the simple interest formula in each of these problems to simplify the process. (With simple interest, the interest is figured on the beginning amount only.)

**EXAMPLE**

Khalil had \$20,000 to invest last year. He invested some of this money at $3\frac{1}{2}$ percent interest and the rest at 8 percent interest. His total earnings in interest, for both of the investments, were \$970. How much did he have invested at each rate?

Think of the \$20,000 as being divvied up into two large containers. Let x represent the amount of money invested at $3\frac{1}{2}$ percent.

The first container has the $x$ dollars in it and a label saying $3\frac{1}{2}$. The second container has a label saying 8 percent and $20,000 - x$ dollars in it. The number $970 is on a sign next to the containers. That's the result of multiplying the mixture (combined) percentage times the total investment of $20,000. You don't need to know the mixture percentage — just the result.

$$3\frac{1}{2} \text{ percent}(x) + 8 \text{ percent}(20,000 - x) = 970$$
$$0.035(x) + 0.08(20,000 - x) = 970$$
$$0.035x + 1,600 - 0.08x = 970$$

Subtract 1,600 from each side and simplify on the left side:

$$-0.045x = -630$$

Dividing each side by $-0.045$, you get

$$x = 14,000$$

That means that $14,000 was invested at $3\frac{1}{2}$ percent and the other $6,000 was invested at 8 percent.

**EXAMPLE**

Kathy wants to withdraw only the interest on her investment each year. She's going to put money into the account and leave it there, just taking the interest earnings. She wants to take out and spend $10,000 each year. If she puts two-thirds of her money where it can earn 5 percent interest and the rest at 7 percent interest, how much should she put at each rate to have the $10,000 spending money?

Let $x$ represent the total amount of money Kathy needs to invest. Using those handy, dandy containers again, the first container has a label of 5 percent and contains $\frac{2}{3}x$ dollars. The second container has 7 percent on its label and has $\frac{1}{3}x$ dollars. The mixture has $10,000; this is the result of the "mixed" percentage and the total amount invested.

$$5\%\left(\frac{2}{3}x\right) + 7\%\left(\frac{1}{3}x\right) = 10,000$$

Change the decimals to fractions and multiply:

$$0.05\left(\frac{2}{3}x\right)+0.07\left(\frac{1}{3}x\right)=10,000$$

$$\frac{1}{30}x+\frac{7}{300}x=10,000$$

Find a common denominator and add the coefficients of $x$:

$$\frac{17}{300}x=10,000$$

Divide each side by $\frac{17}{300}$:

$$x\approx176,470.59$$

Kathy needs over \$176,000 in her investment account. Two-thirds of it, about \$117,647, has to be invested at 5 percent and the rest, about \$58,824, at 7 percent.

## Greening up with money

Money is everyone's favorite topic. When you're combining money and algebra, you have to consider the number of coins or bills and their worth or denomination. Other situations involving money can include admission prices, prices of different pizzas in an order, or any commodity with varying prices.

**EXAMPLE**

Chelsea has five times as many quarters as dimes, three more nickels than dimes, and two fewer than nine times as many pennies as dimes. If she has \$15.03 in coins, how many of them are quarters?

Some containers work here, too. There will be four of them added together, labeled: *dimes*, *quarters*, *nickels*, and *pennies*. Also, on the label, is the value of each coin. Every coin count refers to dimes in this problem, so let the number of dimes be represented by $x$ and compare everything else to it.

The first container would contain dimes; put 0.10 and $x$ on the label. The second container contains quarters; put 0.25 and $5x$ on the label. The third container contains nickels; so put 0.05 and $x+3$ on the label. The fourth container contains pennies; so put 0.01 and $9x-2$ on the label. A mixture container, on the right, has \$15.03 on it.

$$0.10(x)+0.25(5x)+0.05(x+3)+0.01(9x-2)=15.03$$
$$0.10x+1.25x+0.05x+0.15+0.09x-0.02=15.03$$

Simplifying on the left, you get

$1.49x + 0.13 = 15.03.$

Subtracting 0.13:

$1.49x = 14.90$

And, after dividing by 1.49,

$x = 10$

Because $x$ is the number of dimes, there are 10 dimes, five times as many or 50 quarters, three more or 13 nickels and two fewer than nine times as many or 88 pennies. The question was, "How many quarters?" There were 50 quarters; use the other answers to check to see if this comes out correctly.

# Formulating Distance Problems

You travel, I travel, everybody travels, and at some point everybody asks, "Are we there yet?" Algebra can't answer that question for you, but it can help you estimate how long it takes to get there — wherever "there" is.

## Making the distance formula work for you

You've been on a slow boat to China for a couple of days and want to know how far you've come. Or you want to figure out how long it'll take for the rocket to reach Jupiter. Or maybe you want to know how fast a train travels if it gets you from Toronto, Ontario, to Miami, Florida, in 18 hours. The distance = rate · time formula can help you find the answer to all these questions.

**ALGEBRA RULES**

The formula $d = rt$ means the distance traveled is equal to the rate $r$ (the speed) times how long it takes, $t$ (the time). Solving the formula for either the rate or the time, you get: $r = \frac{d}{t}$ and $t = \frac{d}{r}$. Given any two of the values, you can solve for the third. You change the original formula to one that you can use.

**EXAMPLE**

If a plane travels 2,000 miles in 4.8 hours, then what was the average speed of the plane during the trip?

You're looking for the speed or *rate*, *r*, so you use this formula:

$$r = \frac{d}{t}$$

So, plugging in the numbers, $r = \frac{2,000}{4.8} = 416\frac{2}{3}$ miles per hour.

**REMEMBER**

Always be sure that the units are the same: Miles per day and total number of miles go together, but miles per hour and total number of days would take some adjusting.

**EXAMPLE**

How far did Alberto travel in his triathlon if he swam at 2 miles per hour for 30 minutes, bicycled at 25 miles per hour for 45 minutes, and then ran at 6 miles per hour for 6 minutes?

The distance formula $d = rt$ is used three times and the results added together to get the total distance.

You need to change 2 mph for 30 minutes to 2 mph for $\frac{1}{2}$ hour. Then change 25 mph for 45 minutes to 25 mph for $\frac{3}{4}$ hour. Finally, change 6 mph for 6 minutes to 6 mph for $\frac{1}{10}$ hour. All those fractions of hours come from dividing the number of minutes by 60.

$$\left(2 \times \frac{1}{2}\right) + \left(25 \times \frac{3}{4}\right) + \left(6 \times \frac{1}{10}\right) = 1 + \frac{75}{4} + \frac{6}{10}$$

$$= 1 + 18\frac{3}{4} + \frac{3}{5}$$

$$= 19 + \left(\frac{3}{4} + \frac{3}{5}\right)$$

$$= 19 + \left(\frac{15}{20} + \frac{12}{20}\right)$$

$$= 19 + \frac{27}{20}$$

$$= 19 + 1\frac{7}{20}$$

$$= 20\frac{7}{20}$$

Alberto traveled over 20 miles.

## Figuring distance plus distance

A basic distance problem involves one object traveling a certain distance, a second object traveling another distance, and the two distances getting added together.

**EXAMPLE**

Deirdre and Donovan are in love and will be meeting in Kansas City to get married. Deirdre boarded a train at noon traveling due east toward Kansas City. Two hours later, Donovan boarded a train traveling due west, also heading for Kansas City, and going at a rate of speed 20 miles per hour faster than Deirdre. At noon, they were 1,100 miles apart. At 9 p.m., they both arrived in Kansas City. How fast were they traveling?

distance of Deirdre from Kansas City + distance of Donovan from Kansas City = 1,100

(rate × time) + (rate × time) = 1,100

Let the speed (rate) of Deirdre's train be represented by r. Donovan's train was traveling 20 miles per hour faster than Deirdre's, so the speed of Donovan's train is $r + 20$.

Let the time traveled by Deirdre's train be represented by t. Donovan's train left two hours after Deirdre's, so the time traveled by Donovan's train is $t - 2$. Substituting the expressions into the first equation,

$$rt + (r + 20)(t - 2) = 1,100$$

Deirdre left at noon and arrived at 9, so $t = 9$ hours for Deirdre's travels and $t - 2 = 7$ hours for Donovan's. Replacing these values in the equation,

$$r(9) + (r + 20)(7) = 1,100$$

Now distribute the 7:

$$9r + 7r + 140 = 1,100$$

Combine the two terms with r:

$$16r = 960$$

Divide each side by 16:

$$r = 60$$

Deirdre's train is going 60 mph; Donovan's is going $r + 20 = 80$ mph.

## Figuring distance and fuel

My son, Jim, sent me this problem when he was stationed in Afghanistan with the Marines. He was always a whiz at story problems — doing them in his head and not wanting to show any work. He must have been listening to me, because, at the end of his contribution, he added, "Don't forget to show your work!"

**EXAMPLE**

A CH-47 troop-carrying helicopter can travel 300 miles if there aren't any passengers. With a full load of passengers, it can travel 200 miles before running out of fuel. If Camp Tango is 120 miles away from Camp Sierra, can the CH-47 carry a full load of Special Forces members from Tango to Sierra, drop off the troops, and return safely to Tango before running out of fuel? If so, what percentage of fuel will it have left?

I felt a little nervous, working on this problem, with so much at stake. So I took my own advice and drew a picture, tried some scenarios with numbers, and assigned a variable to an amount.

Let $x$ represent the number of gallons of fuel available in the helicopter, and write expressions for the amount used during each part of the operation.

When the helicopter is loaded, it can travel 200 miles on a full tank of fuel. The camps are 120 miles apart, so the helicopter uses $\frac{120}{200}x$ gallons for that part of the trip.

When there are no passengers, the helicopter can travel 300 miles on a full tank. So it uses $\frac{120}{300}x$ gallons for the return flight.

Adding the two amounts together:

$$\frac{120}{200}x + \frac{120}{300}x = \frac{3}{5}x + \frac{2}{5}x = \frac{5}{5}x = x$$

It looks like there's no room for a scenic side-trip. And I haven't figured in the fuel needed for landing and taking off. Hopefully, there's a reserve tank.

# Stirring Things Up with Mixtures

Mixture problems can take on many different forms. There are the traditional types, in which you can actually mix one solution and another, such as water and antifreeze. There are the types in which different solid ingredients are mixed, such as in a salad bowl or candy dish. Another type is where different investments at different interest rates are mixed together. I cover some of these mixture problems in the earlier section, "Finding Money and Interest Interesting." Here I show you a process you can use to solve most mixture problems — very much like I used in the interest and money problems.

Drawing a picture helps with all mixture problems. The same picture can work for all: liquids, solids, and investments. Figure 11-1 shows three sample containers — two added together to get a third (the mixture). In each case, the containers are labeled with the quality and quantity of the contents. These two values get multiplied together before adding. The quality is the strength of the antifreeze or the percentage of the interest or the price of the ingredient. The quantity is the amount in quarts or dollars or pounds. You can use the same picture for the containers in every mixture problem or you can change to bowls or boxes. It doesn't matter — you just want to visualize the way the mixture is going together.

**FIGURE 11-1:** Visualizing containers can help with mixture problems.

A traditional mixture-type problem involves solutions — where you mix water and antifreeze to create a particular strength of antifreeze. When the liquids are mixed, the strengths of the two liquids begin to average out.

EXAMPLE

How many quarts of 80 percent antifreeze have to be added to 8 quarts of 20 percent antifreeze to get a mixture of 60 percent antifreeze?

First, label your containers. The first would be labeled 80 percent on the top and $x$ on the bottom. (I don't know yet how many quarts have to be added.) The second container would be labeled with 20 percent on the top and 8 quarts on the bottom. The third container, which represents the final mixture, would be labeled 60 percent on top and $x + 8$ quarts on the bottom. To solve this, multiply each "quality" or percentage strength of antifreeze times its "quantity" and put these in the equation:

$$80\%(x) \text{ quarts} + 20\%(8) \text{ quarts} = 60\%(x+8) \text{ quarts}$$
$$(0.8)(x) + (0.2)(8) = (0.6)(x+8)$$
$$0.8x + 1.6 = 0.6x + 4.8$$

Subtracting 0.6x from each side and subtracting 1.6 from each side, I get

$$0.2x = 3.2$$

Dividing each side of the equation by 0.2, I get

$$x = 16$$

So 16 quarts of 80 percent antifreeze have to be added.

You can use the liquid mixture rules with salad dressings, mixed drinks, and all sorts of sloshy concoctions.

# Chapter **12**
# Putting Geometry into Story Problems

You can't get away from it: square yards of carpeting, miles per gallon for the car, capacity of the new freezer. In this chapter, I reacquaint you with area, perimeter, and volume. You also see how to deal with those awkward, irregularly shaped objects. It isn't all that important that you memorize the formulas — the main emphasis is on how to use the formula and where to find it when you need it.

## Triangulating a Problem with the Pythagorean Theorem

A wonderful formula to use when working with lengths and triangular situations is the Pythagorean theorem. The Pythagorean theorem is a formula that shows the special relationship between the three sides of a right triangle.

**ALGEBRA RULES**

According to the Pythagorean theorem, if $a$, $b$, and $c$ are the lengths of the sides of a right triangle (like the ones shown in Figure 12-1), and $c$ is the longest side (the hypotenuse), then $a^2 + b^2 = c^2$.

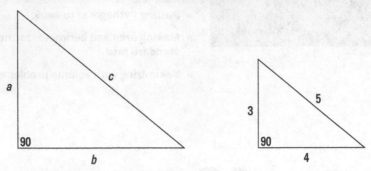

FIGURE 12-1: Triangulating the "right" way.

**EXAMPLE**

A carpenter wants to determine whether a garage doorway has square corners or if it's really leaning to one side. She measures 30 inches from one corner along the bottom of the doorway and makes a mark. She measures 40 inches up along the door frame from the same corner and makes a mark on the side. She then takes a tape measure and measures the distance between the marks; it comes out to be 49 inches.

Find the squares of the measures:

$$30^2 = 900 \qquad 40^2 = 1,600 \qquad 49^2 = 2,401$$

Then, $900 + 1,600 = 2,500 \neq 2,401$. The two smaller squares don't add up to the larger square, so the corner isn't square.

# Being Particular about Perimeter

How long is the running track around the field? What's the distance around the room? How many feet of fencing do you need to go around the pool? The *perimeter* is the distance around the outside of a given figure — the total length of the periphery that borders a region.

In general, the perimeter of a figure is the sum of the lengths of the sides.

## Triangulating triangles

**ALGEBRA RULES**

The perimeter of a triangle is equal to the sum of the measures of the three sides: $P = s_1 + s_2 + s_3$.

**EXAMPLE**

Find the amount of fencing you'll need for a triangular area if the two sides that form a right triangle are 7 yards and 24 yards, and you can't measure the longest side, the hypotenuse, because it's too muddy right now.

Because you have a right triangle, the sum of the squares of 7 and 24 is equal to the square of the longest side:

$$7^2 + 24^2 = 49 + 576 = 625$$

Because 625 is the square of 25, the sides of the area are 7, 24, and 25 yards. Then $P = 7 + 24 + 25 = 56$ yards of fencing needed.

## Squaring up to squares and rectangles

A square is wonderful to work with because you have only one measure to worry about — the length of one side is the same as all the others. A rectangle is a special four-sided figure, too. Figure 12-2 shows a rectangle with square (90-degree) corners, where the opposite sides are the same length.

Rectangle

**FIGURE 12-2:** A shape for rooms, posters, and corrals.

**ALGEBRA RULES**

To find the perimeter of a square or rectangle, use the following formulas:

» The perimeter of a *square* is four times the length of a side: $P = 4s$ (which is easier than adding $s_1 + s_2 + s_3 + s_4$).

» The perimeter of a rectangle is twice the length plus twice the width. Or you can add the length and width together and then multiply that sum by two. These formulas are easier than adding up the four sides: $P = 2l + 2w = 2(l + w)$ or $P = s_1 + s_2 + s_3 + s_4$.

The following examples illustrate using the formulas for perimeter.

**EXAMPLE**

An environmental group is going to search a square mile of prairie to check for toxins in beetles. What is the perimeter of that square mile in feet?

You know that 1 mile is 5,280 feet. So the perimeter is $4 \cdot 5,280 = 21,120$ feet. So, if they want to rope off the area, they need plenty of rope!

**EXAMPLE**

Your new garden is a rectangle measuring 85 feet long by 35 feet wide. How much fencing do you need to enclose it?

What's the perimeter? Add the 85 and 35 together and double it: $2(85 + 35) = 2(120) = 240$ feet of fencing. Of course, this doesn't include a gate — you should probably consider that, too, unless you like jumping hurdles.

## Recycling circles

A circle has a perimeter, but there's a special name for that perimeter: *circumference*. To find the circumference of a circle, all you need is the measure of the radius or the diameter. The radius is the distance from the center of the circle to any point on the circle. If you double the radius, you get the measure of the *diameter*, the distance from one side to the other through the center.

**ALGEBRA RULES**

The formula for *circumference* (distance around the outside of a circle) is $C = 2\pi r = \pi d$ where $r$ is the radius, $d$ is the diameter, and $\pi$ is always about 3.14 or about $\frac{22}{7}$.

**EXAMPLE**

You want to construct a circular garden but you're a member of the waste-not-want-not club. The fencing you want comes in bundles of 50 feet, 100 feet, 150 feet, 200 feet, and so on, so you're going to construct your garden such that it uses every bit of the fencing around the circumference. How can you easily determine the diameter of each garden with respect to the different fencing amounts?

You should rewrite the formula so you can easily determine how wide your circular garden will be if you buy a certain size bundle of fencing to put around it and use all the fencing in the bundle.

Solving for $d$ in the formula $C = \pi d$, divide each side by $\pi$:

$$\frac{C}{\pi} = \frac{\pi d}{\pi}$$

$$\frac{C}{\pi} = d$$

The diameter is equal to the circumference divided by $\pi$.

$$d = \frac{C}{\pi}$$

If the bundle has 50 feet of fencing,

$$d = \frac{50}{3.14} \approx 15.92 \text{ feet across}$$

If the bundle has 100 feet of fencing,

$$d = \frac{100}{3.14} \approx 31.85 \text{ feet across}$$

If the bundle has 200 feet of fencing,

$$d = \frac{200}{3.14} \approx 63.69 \text{ feet across}$$

If you know the dimensions of the lot where you're putting your garden, you can determine which garden will fit.

# Making Room for Area Problems

Area is a measure of how many two-dimensional units (squares) a particular object or surface covers — how much flat space it occupies. Usually, area is given in square inches, square centimeters, square feet, or square miles, and so on.

## Ruminating about rectangles and squares

Rectangles and squares have basically the same area formulas because they both have square corners and the equal lengths on opposite sides. The general procedure here is just to multiply the measure of the length times the measure of the width. The product of two sides that are next to one another is the area.

Most rooms in homes and offices are rectangular in shape. Desks and tables and rugs are usually rectangular, also. This makes it easy to fit furniture and other objects in the room.

ALGEBRA
RULES

The area of a rectangle is its length times its width, and the area of a square is the square of the measure of any side:

Rectangle: $A = lw$

Square: $A = s^2$

EXAMPLE

A garden 85 feet long by 35 feet wide needs some fertilizer. If a bag of fertilizer covers 6 square yards, how many bags of fertilizer do you need?

Note that the measures are different. The garden is measured in feet and the fertilizer coverage is in square yards. Determine how many square feet the garden is. Then convert the fertilizer coverage to square feet per bag.

$$\text{area of garden} = l \times w = 85 \times 35 = 2{,}975 \text{ square feet}$$

Now, how many square feet are there in a square yard? If a yard is equal to 3 feet, then a square yard is 3 feet by 3 feet, so the area is $3^2 = 9$ square feet. There are 9 square feet in a square yard. A bag of fertilizer covers 6 square yards, so that's $6 \times 9 = 54$ square feet per bag.

Divide the 2,975 square feet by 54 square feet per bag:

$$\frac{2{,}975}{54} = 55\frac{5}{54} \approx 55.09 \text{ bags}$$

You can buy 56 bags and have a lot left over or buy 55 bags and skimp a little in some places.

## Taking on triangles

Finding the area of a triangle can be a bit of a challenge. Basically, a triangle's area is half that of an imaginary parallelogram that the triangle fits into. However, it isn't always easy or necessary to find the length and width of this hypothetical parallelogram — you just need a measurement or two from the triangle.

The traditional formula for finding the area of a triangle involves the length of the base, or bottom, and the height, the perpendicular distance from the base up to the *vertex* (the intersection of the other two sides).

**ALGEBRA RULES**

The area of a triangle is equal to half the product of the measure of the base of the triangle, *b*, times the height of the triangle, *h*: $A = \frac{1}{2}bh$.

The base is the length of the bottom that the height is drawn down to. The height is the length from the top angle down perpendicular to the base. The height forms a right angle (90 degrees) with the base. Figure 12-3 shows you a triangle with a height drawn.

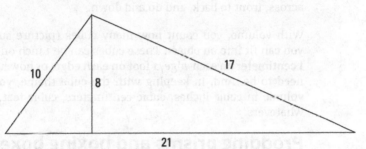

**FIGURE 12-3:** Triangles come in all shapes and sizes.

You use this traditional rule for area when it's possible to make these measurements — when you can draw the height perpendicular to the base and measure both of them.

**EXAMPLE**

Find the area of a triangle 21 feet long with a height of 8 feet. Refer to Figure 12-3 for a sketch of such a triangle.

$$A = \frac{1}{2}(21)(8) = \frac{1}{2}(168) = 84 \text{ square feet}$$

# Rounding up circles

The area of a circle is tied to both the radius of the circle and the value of *π*.

**ALGEBRA RULES**

The formula for the area of a circle is *π* (about 3.14) times the radius squared: $A = \pi r^2$.

EXAMPLE

Find the area of a circular disk that is 50 feet across. First, you need to find the radius. If the circle is 50 feet across, that's the measure of the diameter, all the way across. So the radius is half that or 25 feet. Using the formula to find the area:

$$A = \pi r^2 = \pi \cdot 25^2 = 3.14 \cdot 625 = 1,962.5 \text{ square feet}$$

# Validating with Volume

Area is a two-dimensional figure or representation. It's a flat region. Volume is three-dimensional. To find volume, you measure across, front to back, and up and down.

With volume, you count how many cubes (picture sugar cubes) you can fit into an object. These cubes can be 1 inch on each edge, 1 centimeter on each edge, 1 foot on each edge, or however big they need to be. And, in keeping with the cube theme, you measure volume in cubic inches, cubic centimeters, cubic feet, and cubic whatevers.

## Prodding prisms and boxing boxes

The volume of a rectangular prism, better known as a box, is one of the simplest to find in the world of volume problems. The bottom and top of a prism have exactly the same measurements. The distance from the top to bottom is the same, no matter where you measure, as long as you keep that distance perpendicular to both top and bottom.

ALGEBRA
RULES

The formula for finding the volume of a prism is $V = lwh$, which means that the volume is equal to the product of the length, $l$, times the width, $w$, times the height, $h$.

EXAMPLE

If you're buying a 12-cubic-foot refrigerator, what are the dimensions (how big is it)?

There are an infinite number of ways to multiply three numbers together to get 12. Go through some integers and some fractions.

Try to picture what the refrigerator would look like with these dimensions:

» $12 = 1(1)(12)$. That's 1 foot long, 1 foot wide, and 12 feet tall!

» $12 = 2(1)(6)$. That's 2 feet long, 1 foot wide, and 6 feet tall.

» $12 = 2(3)(2)$. That's 2 feet long, 3 feet wide, and 2 feet tall.

» $12 = 1\frac{1}{2}\left(1\frac{1}{2}\right)\left(5\frac{1}{3}\right)$. That's $1\frac{1}{2}$ feet long, $1\frac{1}{2}$ feet wide, and $5\frac{1}{3}$ feet tall.

Which refrigerator would you want? How tall are you? How far can you reach into the back?

## Cycling cylinders

Cylinders were my brother's favorite shape when he was in the Navy on the aircraft carrier USS *Guadalcanal*. Being the wonderful sister that I am, I would send him chocolate chip cookies that fit exactly into a 3-pound coffee can. Imagine a stack of chocolate chip cookies coming to you every couple of weeks. Was he ever popular on *that* ship!

ALGEBRA
RULES

The formula for the volume of a cylinder is $V = \pi r^2 h$. The volume is equal to $\pi$ times the radius (halfway across a circle) squared times the height.

To find the volume of a cylinder, you need the radius of the top and bottom, and you need the height. This formula tells you how many cubes will fit in the cylinder — like putting square pegs in a round hole — just trim them a bit.

EXAMPLE

Find the volume of an above-ground swimming pool that has a radius of 12 feet and a height of 4 feet.

Using the formula for the volume of a cylinder:

$V = \pi r^2 h = \pi(12^2)(4) = \pi(576) \approx 3.14(576) = 1,808.64$ cubic feet of water

## Pointing to pyramids and cones

A pyramid is an easy thing to describe because everyone has a mental picture of what a pyramid looks like. Technically, a pyramid is an object with a base (bottom) and triangles coming up from each side of the base to meet at a point. The base can be any polygon: a triangle, rectangle, square, and so on.

Cones are also very familiar. You see those orange shapes along the road in construction zones, and you hold them very carefully when they're full of drippy ice cream. I put these two figures together, because their volume formulas are so similar. Both volume formulas are essentially one-third of their height times the area of their base.

**ALGEBRA RULES**

The formula for the volume of a pyramid is $V = \frac{1}{3}(\text{area of base}) \cdot h$.

The formula for the volume of a cone is $V = \frac{1}{3}\pi r^2 h$.

**EXAMPLE**

Find the original volume of the Great Pyramid, which originally had a square base with each side measuring 756 feet and a height of 480 feet.

The base is a square, so the area of the base is $s^2$:

$$V = \frac{1}{3}s^2 \cdot h = \frac{1}{3}(756)^2 \cdot 480 = 91,445,760 \text{ cubic feet}$$

**EXAMPLE**

What is the volume of a cone-shaped tent that has a diameter of 18 feet and a height of 20 feet?

If the diameter is 18 feet, then the radius is 9 feet:

$$V = \frac{1}{3}\pi r^2 h = \frac{1}{3}\pi(9)^2 \cdot 20 = 540\pi \approx 1,696 \text{ cubic feet}$$

# Chapter **13**

# Grappling with Graphing

I n this chapter, I present the basics for working with lines and their equations. You find lines determined by two points and then other lines determined by a slope and a point. You see lines that meet and lines that avoid one another forever.

I also throw you a curve or two! Circles and parabolas are the most recognizable of the algebraic curves and have the most respectable equations. The basics for drawing these curves are found here.

## Preparing to Graph a Line

A straight *line* is the set of all the points on a graph that satisfy a linear equation. When any two points on a line are chosen, the *slope* of the segment between those two points is always the same number.

To graph a line, you need only two points. A rule in geometry says that only one line can go through two particular points. Even though only two points are needed to graph a line, it's usually a good idea to graph at least three points to be sure that you graphed the line correctly.

An equation whose graph is a straight line is said to be *linear*. A linear equation has a standard form of $ax + by = c$, where $x$ and $y$

are variables and $a$, $b$, and $c$ are real numbers. A point $(x, y)$ lies on the line if the $x$ and $y$ make the equation true. When graphing a line, you can find some pairs of numbers that make the equation true and then connect them. Connect the dots!

Graphing lines from their equations just takes finding enough points on the line to convince you that you've drawn the graph correctly.

Find a point on the line $x - y = 3$.

**EXAMPLE**

1. **Choose a random value for one of the variables, either $x$ or $y$.**

   To make the arithmetic easy for yourself, pick a large enough number so that, when you subtract $y$ from that number, you get a positive 3. In $x - y = 3$, you can let $x = 8$, so $8 - y = 3$.

2. **Solve for the value of the other variable.**

   Subtract 8 from each side to get $-y = -5$.

   Multiply each side by $-1$ to get $y = 5$.

3. **Write an ordered pair for the coordinates of the point.**

   You chose 8 for $x$ and solved to get $y = 5$, so your first ordered pair is (8, 5).

You can find more ordered pairs by choosing another number to substitute for either $x$ or $y$.

Find a point that lies on the line $2x + 3y = 12$.

**EXAMPLE**

1. **Solve the equation for one of the variables.**

   Solving for $y$ in the sample problem, $2x + 3y = 12$, you get

   $$3y = 12 - 2x$$
   $$y = \frac{12 - 2x}{3}$$

   With multipliers involved, you often get a fraction.

2. **Choose a value for the other variable and solve the equation.**

   Try to pick values so that the result in the numerator is divisible by the 3 in the denominator — giving you an integer.

144

For example, let $x = 3$. Solving the equation:

$$y = \frac{12 - 3 \cdot 3}{2} = \frac{6}{3} = 2$$

So, the point (3, 2) lies on the line.

# Incorporating Intercepts

An *intercept* of a line is a point where the line crosses an axis. Unless a line is vertical or horizontal, it crosses both the $x$ and $y$ axes, so it has two intercepts: an $x$-intercept and a $y$-intercept. Horizontal lines have just a $y$-intercept, and vertical lines have just an $x$-intercept. The exceptions are when the horizontal line is actually the $x$-axis or the vertical line is the $y$-axis. Intercepts are quick and easy to find and can be a big help when graphing.

**ALGEBRA RULES**

The $x$-intercept of a line is where the line crosses the $x$-axis. To find the $x$-intercept, let the $y$ in the equation equal 0 and solve for $x$.

**EXAMPLE**

Find the $x$-intercept of the line $4x - 7y = 8$.

First, let $y = 0$ in the equation. Then:

$$4x - 0 = 8$$
$$4x = 8$$
$$x = 2$$

The $x$-intercept of the line is (2, 0). The line goes through the $x$-axis at that point.

**ALGEBRA RULES**

The $y$-intercept of a line is where the line crosses the $y$-axis. To find the $y$-intercept, let the $x$ in the equation equal 0 and solve for $y$.

**EXAMPLE**

Find the $y$-intercept of the line $3x - 7y = 28$.

First, let $x = 0$ in the equation. Then:

$$0 - 7y = 28$$
$$-7y = 28$$
$$y = -4$$

The $y$-intercept of the line is $(0, -4)$.

**TIP**

As long as you're careful when graphing the $x$- and $y$-intercepts and get them on the correct axes, the intercepts are often all you need to graph a line.

# Sliding the Slippery Slope

The slope of a line is a number that describes the steepness and direction of the graph of the line. The slope is a positive number if the line moves upward from left to right; the slope is a negative number if the line moves downward from left to right. The steeper the line, the greater the absolute value of the slope (the farther the number is from 0).

Knowing the slope of a line beforehand helps you graph the line. You can find a point on the line and then use the slope and that point to graph it. A line with a slope of 6 goes up steeply. If you know what the line should look like (that is, whether it should go up or down) — information you get from the slope — you'll have an easier time graphing it correctly.

Figure 13-1 shows some lines with their slopes. The lines are all going through the origin just for convenience.

What about a horizontal line — one that doesn't go upward or downward? A horizontal line has a 0 slope. A vertical line has no slope; the slope of a vertical line (it's so steep) is undefined.

**TIP**

One way of referring to the slope, when it's written as a fraction, is rise over run. If the slope is $\frac{3}{2}$, it means that for every 2 units the line runs left to right along the $x$-axis, it rises 3 units along the $y$-axis. A slope of $\frac{-1}{8}$ indicates that as the line runs 8 units horizontally, parallel to the $x$-axis left to right, it drops (negative rise) 1 unit vertically.

## Computing slope

If you know two points on a line, you can compute the number representing the slope of the line.

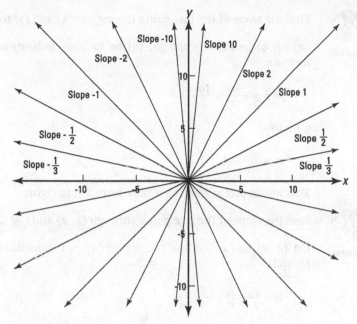

**FIGURE 13-1:** Pick a line — see its slope.

The slope of a line, denoted by the small letter $m$, is found when you know the coordinates of two points on the line, $(x_1, y_1)$ and $(x_2, y_2)$:

$$m = \frac{y_2 - y_1}{x_2 - x_1}$$

Subscripts are used here to identify which is the first point and which is the second point. There's no rule as to which is which; you can name the points any way you want. It's just a good idea to identify them to keep things in order. Reversing the points in the formula gives you the same slope (when you subtract in the opposite order):

$$m = \frac{y_1 - y_2}{x_1 - x_2}$$

You just can't mix them and do $(y_1 - y_2)$ over $(x_2 - x_1)$.

Now, you can see how to compute slope with the following examples.

Find the slope of the line going through (3, 4) and (2, 10).

Let (3, 4) be $(x_1, y_1)$ and (2, 10) be $(x_2, y_2)$. Substitute into the formula:

$$m = \frac{y_2 - y_1}{x_2 - x_1} = \frac{10 - 4}{2 - 3}$$

Simplify:

$$m = \frac{6}{-1} = -6$$

This line is pretty steep as it falls from left to right.

EXAMPLE

Find the slope of the line going through (4, 2) and (–6, 2).

Let (4, 2) be $(x_1, y_1)$ and (–6, 2) be $(x_2, y_2)$. Substitute into the formula:

$$m = \frac{y_2 - y_1}{x_2 - x_1} = \frac{2 - 2}{-6 - 4}$$

Simplify:

$$m = \frac{0}{-10} = 0$$

These points are both 2 units above the x-axis and determine a horizontal line. That's why the slope is 0.

EXAMPLE

Find the slope of the line going through (2, 4) and (2, –6).

Let (2, 4) be $(x_1, y_1)$ and (2, –6) be $(x_2, y_2)$. Substitute into the formula:

$$m = \frac{y_2 - y_1}{x_2 - x_1} = \frac{-6 - 4}{2 - 2}$$

Simplify:

$$m = \frac{-10}{0}$$

Oops! You can't divide by 0. There is no such number. The slope doesn't exist or is undefined. These two points are on a vertical line.

Watch out for these common errors when working with the slope formula:

>> **Be sure that you subtract the *y* values on the top of the division formula.** A common error is to subtract the *x* values on the top.

>> **Be sure to keep the numbers in the same order when you subtract.** Decide which point is first and which point is second. Then take the second *y* minus the first *y* and the second *x* minus the first *x*. Don't do the top subtraction in a different order from the bottom.

## Combining slope and intercept

An equation of a single line can take many forms. Just as you can solve for one variable or another in a formula, you can solve for one of the variables in the equation of a line. This change of format can help you find the points to graph the line or find the slope of a line.

A common and popular form of the equation of a line is the *slope-intercept form*. It's given this name because the slope of the line and the *y*-intercept of the line are obvious on sight. When a line is written $6x + 3y = 5$, you can find points by plugging in numbers for *x* or *y* and solving for the other coordinate. But, by using methods for solving linear equations (see Chapter 6), the same equation can be written $y = -2x + \dfrac{5}{3}$, which tells you that the slope is $-2$ and the place where the line crosses the *y*-axis (the *y*-intercept) is $\left(0, \dfrac{5}{3}\right)$.

Where *y* and *x* represent coordinates of a point on the line, *m* is the slope of the line, and *b* is the *y*-intercept of the line, the slope-intercept form is $y = mx + b$.

In every case shown next, the equation is written in the slope-intercept form. The coefficient of *x* is the slope of the line and the constant gives the *y*-intercept.

>> $y = 2x + 3$: The slope is 2; the *y*-intercept is $(0, 3)$.

>> $y = \dfrac{1}{3}x - 2$: The slope is $\dfrac{1}{3}$; the *y*-intercept is $(0, -2)$.

>> $y = 7$: The slope is 0; the $y$-intercept is (0, 7). You can read this equation as being $y = 0 \cdot x + 7$.

## Creating the slope-intercept form

If the equation of the line isn't already in the slope-intercept form, solving for $y$ changes the equation to slope-intercept form.

Put the equation $5x - 2y = 10$ in slope-intercept form.

**EXAMPLE**

**1.** **Get the $y$ term by itself on the left.**

Subtract $5x$ from each side to get the $y$ term alone:

$-2y = -5x + 10$

**2.** **Solve for $y$.**

Divide each side by $-2$ and simplify the two terms on the right:

$$\frac{-2y}{-2} = \frac{(-5x+10)}{-2}$$

$$y = \frac{-5x}{-2} + \frac{10}{-2}$$

$$y = \frac{5}{2}x - 5$$

The slope is $\frac{5}{2}$ and the $y$-intercept is at $(0, -5)$.

## Graphing with slope-intercept

One advantage to having an equation in the slope-intercept form is that graphing the line can be a fairly quick task, as the following example shows.

Graph $y = \frac{3}{2}x + 1$.

**EXAMPLE**

The slope of this line is $\frac{3}{2}$ and the $y$-intercept is the point (0, 1). First, graph the $y$-intercept (see Figure 13-2). Then use the rise-over-run interpretation of slope to count spaces to another point on the line. To do this, do the run, or bottom, movement first. In this sketch, move 2 units to the right of (0, 1). From there, rise (or go up) 3 units, which should get you to (2, 4).

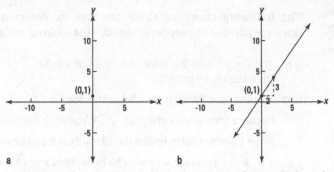

FIGURE 13-2: The y-intercept is located; use run and rise to find another point.

It's sort of like going on a treasure hunt: "Two steps to the east; three steps to the north; now dig in!" Only our "dig in" is to put a point there and connect that point with the starting point — the intercept. Look at the right-hand side (the b side) of Figure 13-2 to see how it's done.

# Making Parallel and Perpendicular Lines Toe the Line

The slope of a line gives you information about a particular characteristic of the line. It tells you if it's steep or flat and if it's rising or falling as you read from left to right. The slope of a line can also tell you if one line is parallel or perpendicular to another line.

Parallel lines never touch. They're always the same distance apart and never share a common point. They have the same slope.

Perpendicular lines form a 90-degree angle (a *right angle*) where they cross. They have slopes that are negative reciprocals of one another. For example, the x-axis and y-axis are perpendicular lines.

**ALGEBRA RULES**

If line $l_1$ has a slope of $m_1$, and if line $l_2$ has a slope of $m_2$, then the lines are parallel if $m_1 = m_2$. If line $l_1$ has a slope of $m_1$, and if line $l_2$ has a slope of $m_2$, then the lines are perpendicular if $m_1 = -\frac{1}{m_2}$ or if they are horizontal or vertical lines.

The following examples show you how to determine whether lines are parallel or perpendicular by just looking at their slopes:

≫ The line $y = 3x + 2$ is parallel to the line $y = 3x - 7$ because their slopes are both 3.

≫ The line $3x + 2y = 8$ is parallel to the line $6x + 4y = 7$ because their slopes are both $\frac{-3}{2}$. Write each line in the slope-intercept form to see this: $3x + 2y = 8$ can be written $y = \frac{-3}{2}x + 4$ and $6x + 4y = 7$ can be written $y = \frac{-3}{2}x + \frac{7}{4}$.

≫ The line $y = \frac{3}{4}x + 5$ is perpendicular to the line $y = \frac{-4}{3}x + 6$ because their slopes are negative reciprocals of one another.

≫ The line $y = -3x + 4$ is perpendicular to the line $y = \frac{1}{3}x - 8$ because their slopes are negative reciprocals of one another.

## Criss-Crossing Lines

If two lines *intersect*, or cross one another, then they intersect exactly once and only once. The place they cross is the point of intersection and that common point is the only one both lines share. Careful graphing can sometimes help you to find the point of intersection.

The point (5, 1) is the point of intersection of the two lines $x + y = 6$ and $2x - y = 9$ because the coordinates make each equation true:

≫ If $x + y = 6$, then substituting the values $x = 5$ and $y = 1$ give you $5 + 1 = 6$, which is true.

≫ If $2x - y = 9$, then substituting the values $x = 5$ and $y = 1$ give $2 \cdot 5 - 1 = 10 - 1 = 9$, which is also true.

This is the only point that works for both the lines.

One way to find the intersection of two lines is to graph both lines (very carefully) and observe where they cross. This technique is not very helpful when the intersection has fractional coordinates, though.

Another way to find the point where two lines intersect is to use a technique called *substitution* — you substitute the $y$ value from one equation for the $y$ value in the other equation and then solve for $x$. Because you're looking for the place where $x$ and $y$ of each line are the same — that's where they intersect — then you can write the equation $y = y$, meaning that the $y$ from the first line is equal to the $y$ from the second line. Replace the $y$'s with what they're equal to in each equation, and solve for the value of $x$ that works.

Find the intersection of the lines $3x - y = 5$ and $x + y = -1$.

**EXAMPLE**

**1.** **Put each equation in the slope-intercept form, which is a way of solving each equation for $y$.**

$3x - y = 5$ is written as $y = 3x - 5$, and $x + y = -1$ is written as $y = -x - 1$. (The lines are not parallel, and their slopes are different, so there will be a point of intersection.)

**2.** **Set the $y$ points equal and solve.**

From $y = 3x - 5$ and $y = -x - 1$, you substitute what $y$ is equal to in the first equation with the $y$ in the second equation:
$3x - 5 = -x - 1$.

**3.** **Solve for the value of $x$.**

Add $x$ to each side and add 5 to each side:

$$3x + x - 5 + 5 = -x + x - 1 + 5$$
$$4x = 4$$
$$x = 1$$

Substitute that 1 for $x$ into either equation to find that $y = -2$. The lines intersect at the point $(1, -2)$.

# Turning the Curve with Curves

A circle is a most recognizable shape. A circle is basically all the points that are a set distance from the point called the circle's *center*. A parabola isn't quite as recognizable as a circle, but it's represented by a quadratic equation and fairly easy to graph.

# Going around in circles with a circular graph

An example of an equation of a circle is $x^2 + y^2 = 25$. The circle representing this equation goes through an infinite number of points. Here are just some of those points:

| | | | | |
|---|---|---|---|---|
| (0, 5) | (0, -5) | (5, 0) | (-5, 0) | (3, 4) |
| (4, 3) | (4, -3) | (-3, 4) | (-3, -4) | (-4, -3) |

I haven't finished all the possible points with integer coordinates, let alone points with fractional coordinates, such as $\left(\dfrac{25}{13} \cdot \dfrac{60}{13}\right)$.

**TIP**

When graphing an equation, you don't expect to find all the points. You just want to find enough points to help you sketch in all the others without naming them.

# Putting up with parabolas

Parabolas are nice, U-shaped curves. They're the graphs of quadratic equations where either an $x$ term is squared or a $y$ term is squared, but not both are squared at the same time. Parabolas have a highest point or a lowest point (or the farthest left point or the farthest right point) called the *vertex*.

## Trying out the basic parabola

My favorite example of a parabola is $y = x^2$, the basic parabola. Figure 13-3 shows a graph of this formula. This equation says that the $y$-coordinate of every point on the parabola is the square of the $x$-coordinate.

The vertex of the parabola in Figure 13-3 is at the origin, (0, 0), and the graph curves upward.

You can make this parabola steeper or flatter by multiplying the $x^2$ by certain numbers. If you multiply the squared term by numbers bigger than 1, it makes the parabola steeper. If you multiply by numbers between 0 and 1 (which are proper fractions), it makes the parabola flatter.

You can make the parabola open downward by multiplying the $x^2$ by a negative number, and make it steeper or flatter than the basic parabola — in a downward direction.

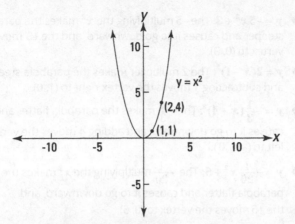

FIGURE 13-3: The simplest parabola.

## Putting the vertex on an axis

The basic parabola, $y = x^2$, can be slid around — left, right, up, down — placing the vertex somewhere else on an axis and not changing the general shape.

If you change the basic equation by adding a constant number to the $x^2$ — such as $y = x^2 + 3$, $y = x^2 + 8$, $y = x^2 - 5$, or $y = x^2 - 1$ — then the parabola moves up and down the $y$-axis. Note that adding a negative number is also part of this rule. These manipulations help make a parabola fit the model of a certain situation.

If you change the basic parabolic equation by adding a number to the $x$ first and then squaring the expression — such as $y = (x + 3)^2$, $y = (x + 8)^2$, $y = (x - 5)^2$, or $y = (x - 1)^2$ — you move the graph to the left or right of where the basic parabola lies. Using +3, as in the equation $y = (x + 3)^2$, moves the graph to the left, and using –3, as in the equation $y = (x - 3)^2$, moves the graph to the right. It's the opposite of what you might expect, but it works this way consistently.

The following equations and their graphs are shown in Figure 13-4:

» $y = 3x^2 - 2$: The 3 multiplying the $x^2$ makes the parabola steeper, and the –2 moves the vertex down to $(0, -2)$.

» $y = \frac{1}{4}x^2 + 1$: The $\frac{1}{4}$ multiplying the $x^2$ makes the parabola flatter, and the +1 moves the vertex up to $(0, 1)$.

» $y = -5x^2 + 3$: The −5 multiplying the $x^2$ makes the parabola steeper and causes it to go downward, and the +3 moves the vertex to (0, 3).

» $y = 2(x - 1)^2$: The 2 multiplier makes the parabola steeper, and subtracting 1 moves the vertex right to (1, 0).

» $y = -\frac{1}{3}(x + 4)^2$: The $-\frac{1}{3}$ makes the parabola flatter and causes it to go downward, and adding 4 moves the vertex left to (−4, 0).

» $y = -\frac{1}{20}x^2 + 5$: The $-\frac{1}{20}$ multiplying the $x^2$ makes the parabola flatter and causes it to go downward, and the +5 moves the vertex to (0, 5).

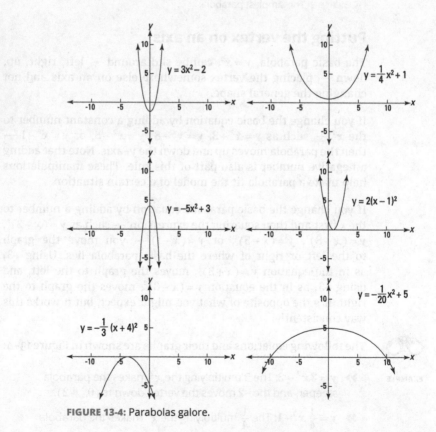

**FIGURE 13-4:** Parabolas galore.

# Chapter **14**

# Ten Warning Signs of Algebraic Pitfalls

S o much algebra is done in the world, so the sheer number of people who use algebra means that a large number of errors are unavoidable. Some errors occur because that error seems to be an easier way to do the problem. Not right, but easier — the path of least resistance. The main errors in algebra occur while performing expanding–type operations: distributing, squaring binomials, breaking up fractions, or raising to powers. The other big problem area is in dealing with negatives. Watch out for those negative vibes.

## Including the Middle Term

A squared binomial has three terms in the answer. The term that often gets left out is the middle term: the part you get when multiplying the two outer terms together and the two inner terms together and finding their sum. The error occurs when just the first and last separate terms are squared, and the middle term is just forgotten.

| Right | Wrong |
|-------|-------|
| $(a+b)^2 = a^2 + 2ab + b^2$ | $(a+b)^2 \neq a^2 + b^2$ |

# Keeping Distributions Fair

Distributing a number or a negative sign over two or more terms in parentheses can cause problems if you forget to distribute the outside value over every single term in the parentheses. The errors come in when you stop multiplying the terms in the parentheses before you get to the end.

| Right | Wrong |
|-------|-------|
| $x - 2(y + z - w) = x - 2y - 2z + 2w$ | $x - 2(y + z - w) \neq x - 2y + z - w$ |

# Creating Two Fractions from One

Splitting a fraction into several smaller pieces is all right as long as each piece has a term from the numerator (top) and the entire denominator (bottom). You can't split up the denominator.

| Right | Wrong |
|-------|-------|
| $\dfrac{x+y}{a+b} = \dfrac{x}{a+b} + \dfrac{y}{a+b}$ | $\dfrac{x+y}{a+b} \neq \dfrac{x}{a} + \dfrac{y}{b}$ |

# Restructuring Radicals

If the expression under a radical has values multiplied together or divided, then the radical can be split up into radicals that multiply or divide. You can't split up addition or subtraction, however, under a radical.

| Right | Wrong |
|-------|-------|
| $\sqrt{a^2 + b^2} = \sqrt{a^2 + b^2}$ | $\sqrt{a^2 + b^2} \neq \sqrt{a^2} + \sqrt{b^2}$ |

*Note:* The radical expression is unchanged, because the sum has to be performed before applying the radical operation.

# Including the Negative (or Not)

The order of operations instructs you to raise the expression to a power before you add or subtract. A negative in front of a term acts the same as subtracting, so the subtracting has to be done last. If you want the negative raised to the power, too, then include it in parentheses with the rest of the value.

| Right | Wrong |
|-------|-------|
| $-3^2 = -9$ | $-3^2 \neq 9$ |
| $(-3)^2 = 9$ | |

# Making Exponents Fractional

A fractional exponent has the power on the top of the fraction and the root on the bottom.

**REMEMBER**

When writing $\sqrt{x}$ as a term with a fractional exponent, $\sqrt{x} = x^{\frac{1}{2}}$. A fractional exponent indicates that there's a radical involved in the expression. The two in the fractional exponent is on the bottom — the root always is the bottom number.

| Right | Wrong |
|-------|-------|
| $\sqrt[5]{x^3} = x^{\frac{3}{5}}$ | $\sqrt[5]{x^3} \neq x^{\frac{5}{3}}$ |

# Keeping Bases the Same

When you're multiplying numbers with exponents, and those numbers have the same base, you add the exponents and leave the base as it is. The bases never get multiplied together.

| Right | Wrong |
|-------|-------|
| $2^3 \cdot 2^4 = 2^7$ | $2^3 \cdot 2^4 \neq 4^7$ |

# Powering Up a Power

To raise a value that has a power (exponent) to another power, multiply the exponents to raise the whole term to a new power. Don't raise the exponent itself to a power — it's the base that's being raised, not the exponent.

| Right | Wrong |
|---|---|
| $(x^2)^4 = x^8$ | $(x^2)^4 \neq x^{16}$ |

# Making Reasonable Reductions

When reducing fractions with a numerator that has more than one term separated by addition or subtraction, then whatever you're reducing the fraction by has to divide every single term evenly in both the numerator and the denominator.

| Right | Wrong |
|---|---|
| $\dfrac{(4+6x)}{4} = \dfrac{(2+3x)}{2}$ | $\dfrac{(4+6x)}{4} \neq \dfrac{(2+6x)}{2}$ |

# Catching All the Negative Exponents

When changing fractions to equivalent expressions with negative exponents, give every single factor in the denominator a negative exponent.

| Right | Wrong |
|---|---|
| $\dfrac{1}{2ab^2} = 2^{-1}a^{-1}b^{-2}$ | $\dfrac{1}{2ab^2} \neq 2a^{-1}b^{-2}$ |

# Index

## Symbols
| | (absolute value), 9
≈ (approximately equal), 9
{ } (braces), 10
[ ] (brackets), 10
÷ (division), 9
— (fraction line), 10
≥ (greater than or equal to), 104
> (greater than), 104
≤ (less than or equal to), 104
< (less than), 104
– (minus or negative number), 9
\* (multiplication), 8
· (multiplication), 8
× (multiplication), 8
( ) (parentheses grouping), 10
π (pi), 9, 136–137, 139, 141
+ (plus or positive number), 9
√ (square root or radical), 9

## A
absolute value equations and inequalities, 115–119
absolute value symbol (| |), 9
addition
  exponents (powers), 17–18
  numbers in inequalities, 107
signed numbers, 10–11
symbol for (+), 9
algebra I
  about, 1–5
  absolute value and inequalities, 115–119
  application problems, 121–131
  cubic equations, 87–92
  decimals, 14–16
  distributing. See distributing
  exponents. See exponent (power)
  factoring. See factoring
  fractions. See fractions
  geometry, 133–142
  graphing. See graphing
  grouping. See grouping
  inequalities. See inequality
  linear expressions, 8, 33, 109. See also linear equation
  number systems, 6–7, 10–14, 84–85
  order of operations, 25–27, 57–58
  powers. See power (exponent)
  quadratic expressions, 33, 36–41. See also entries beginning with quadratic
  radicals. See radical expression

multiplication. *See also* exponent (power); factoring
  in inequalities, 107–108
  in linear equations, 60–61, 65–66
  signed numbers, 12–14
  symbols for (×/*/·), 9
multiplication property of zero (MPZ), 74
multiplier, 9

# N

*n* (in exponential expressions), 18
natural number, 6
negative exponent, 18, 20–21, 29–30, 160
negative number, 9–14, 159
nested expression, 64
number systems
  composite numbers, 7
  counting numbers, 6
  imaginary numbers, 84–85
  integers, 6
  irrational numbers, 7
  natural numbers, 6
  prime numbers, 7
  rational numbers, 6
  real numbers, 6
  signed numbers, 10–14
  whole numbers, 6
number-line graph, 105–106
numerator, 10

# O

operations
  addition, 9–11, 17–18, 107
  defined, 8
  division. *See* division
  inequalities, relationship to, 106–108
  multiplication. *See* multiplication
  order of, 25–27, 57–58. *See also* grouping
  radical expression. *See* radical expression
  signed numbers, 10–14
  subtraction, 9, 12, 17–18, 107
  symbols for, 8–9
order of operations, 25–27, 57–58. *See also* grouping

# P

parabola, 154–156
parallel line, 151–152
parentheses symbol ( ), 10
perfect cube, 43
perfect square, 42
perimeter, 134–137
perpendicular line, 151–152
pi ($\pi$), 9, 136–137, 139, 141
plus sign (+), 9
polynomial, 32, 51–54, 93–95. *See also* binomial; trinomial
positive number, 9–14
power (exponent)

## S

## T

# About the Author

**Mary Jane Sterling** has been an educator since graduating from college. Teaching at the junior high, high school, and college levels, she has had the full span of experiences and opportunities to determine how best to explain how mathematics works. She has been teaching at Bradley University in Peoria, Illinois, for the past 30 years. She is also the author of *Algebra I For Dummies,* 2nd Edition; *Algebra II For Dummies; Trigonometry For Dummies; Math Word Problems For Dummies; Business Math For Dummies;* and *Linear Algebra For Dummies.*

# Publisher's Acknowledgments

**Project Editor:** Elizabeth Kuball

**Senior Acquisitions Editor:**
Lindsay Sandman Lefevere

**Copy Editor:** Elizabeth Kuball

**Assistant Editor:**
Erin Calligan Mooney

**Editorial Program Coordinator:**
Joe Niesen

**Technical Editors:** Tony Bedenikovic,
Michael McAsey

**Senior Editorial Manager:**
Jennifer Ehrlich

**Editorial Supervisor and Reprint
Editor:** Carmen Krikorian

**Editorial Assistants:** Rachelle Amick,
Jennette ElNaggar

**Senior Editorial Assistant:**
David Lutton

**Production Editor:** Siddique Shaik

**Cover Photos:** © Snvv18870020330/
Shutterstock

# Leverage the power

*Dummies* is the global leader in the reference category and one of the most trusted and highly regarded brands in the world. No longer just focused on books, customers now have access to the dummies content they need in the format they want. Together we'll craft a solution that engages your customers, stands out from the competition, and helps you meet your goals.

## Advertising & Sponsorships

Connect with an engaged audience on a powerful multimedia site, and position your message alongside expert how-to content. Dummies.com is a one-stop shop for free, online information and know-how curated by a team of experts.

- Targeted ads
- Video
- Email Marketing

- Microsites
- Sweepstakes sponsorship

**20** **MILLION** PAGE VIEWS EVERY SINGLE MONTH

**15** MILLION **UNIQUE** VISITORS PER MONTH

**43%** OF ALL VISITORS ACCESS THE SITE VIA THEIR MOBILE DEVICES

**700,000** NEWSLETTER SUBSCRIPTIONS TO THE INBOXES OF
*300,000* UNIQUE INDIVIDUALS EVERY WEEK

# of dummies

## Custom Publishing

Reach a global audience in any language by creating a solution that will differentiate you from competitors, amplify your message, and encourage customers to make a buying decision.

- Apps
- Books
- eBooks
- Video
- Audio
- Webinars

## Brand Licensing & Content

Leverage the strength of the world's most popular reference brand to reach new audiences and channels of distribution.

For more information, visit **dummies.com/biz**

# PERSONAL ENRICHMENT

**Staying Sharp**
9781119187790
USA $26.00
CAN $31.99
UK £19.99

**Facebook**
9781119179030
USA $21.99
CAN $25.99
UK £16.99

**Guitar**
9781119293354
USA $24.99
CAN $29.99
UK £17.99

**Investing**
9781119293347
USA $22.99
CAN $27.99
UK £16.99

**Beekeeping**
9781119310068
USA $22.99
CAN $27.99
UK £16.99

**Digital Photography**
9781119235606
USA $24.99
CAN $29.99
UK £17.99

**Meditation**
9781119251163
USA $24.99
CAN $29.99
UK £17.99

**Pregnancy**
9781119235491
USA $26.99
CAN $31.99
UK £19.99

**Samsung Galaxy S7**
9781119279952
USA $24.99
CAN $29.99
UK £17.99

**iPhone**
9781119283133
USA $24.99
CAN $29.99
UK £17.99

**Crocheting**
9781119287117
USA $24.99
CAN $29.99
UK £16.99

**Nutrition**
9781119130246
USA $22.99
CAN $27.99
UK £16.99

# PROFESSIONAL DEVELOPMENT

**Windows 10**
9781119311041
USA $24.99
CAN $29.99
UK £17.99

**AutoCAD**
9781119255796
USA $39.99
CAN $47.99
UK £27.99

**Excel 2016**
9781119293439
USA $26.99
CAN $31.99
UK £19.99

**QuickBooks 2017**
9781119281467
USA $26.99
CAN $31.99
UK £19.99

**macOS Sierra**
9781119280651
USA $29.99
CAN $35.99
UK £21.99

**LinkedIn**
9781119251132
USA $24.99
CAN $29.99
UK £17.99

**Windows 10**
9781119310563
USA $34.00
CAN $41.99
UK £24.99

**SharePoint 2016**
9781119181705
USA $29.99
CAN $35.99
UK £21.99

**Fundamental Analysis**
9781119263593
USA $26.99
CAN $31.99
UK £19.99

**Networking**
9781119257769
USA $29.99
CAN $35.99
UK £21.99

**Office 2016**
9781119293477
USA $26.99
CAN $31.99
UK £19.99

**Office 365**
9781119265313
USA $24.99
CAN $29.99
UK £17.99

**Salesforce.com**
9781119239314
USA $29.99
CAN $35.99
UK £21.99

**Coding**
9781119293323
USA $29.99
CAN $35.99
UK £21.99

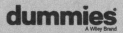